# A KEY TO THE SKULLS
# OF
# NORTH AMERICAN MAMMALS


**Dr. Monte L. Thies**

Department of Biological Sciences
Sam Houston State University
Huntsville, Texas


**FOURTH EDITION**

**Kendall Hunt**
publishing company

Cover image © Shutterstock.com

**Kendall Hunt**
publishing company

www.kendallhunt.com
*Send all inquiries to*:
4050 Westmark Drive
Dubuque, IA  52004-1840

# TABLE OF CONTENTS

# INTRODUCTION TO THE FOURTH EDITION

Since the first edition, this manual has served as both a reference and tool for working in and teaching mammalogy.  However, with continuing revision of the taxonomy of North American mammals, the third edition contains out-of-date information.  My intention with the development of this edition, as was the case with past editions, is to provide an up-to-date comprehensive treatment of all mammal genera, both native and introduced (at least those most likely to be found in the wild), occurring in North America north of Mexico, inclusive of marine mammals found in coastal waters.

Primary changes to the fourth edition include:
- A general reorganization of orders to more accurately represent current associations among higher taxonomic groupings
- Order Cingulata – updated Ordinal name from Xenarthra
- Order Primates – inclusion of the Order and addition of the Families Hominidae (*Homo*) and Cercopithecidae (*Macaca* and *Chlorocebus*)
- Order Rodentia – Family Cricetidae separated from Muridae; elevation of the subgenera of *Spermophilus* to separate genera of *Poliocitellus*, *Otospermophilus*, *Callospermophilus*, *Xerospermophilus*, *Urocitellus*, and *Ictodomys*; reversion of *Pappogeomys* to *Cratogeomys*; elevating subgenus *Myctomys* from *Synaptomys* and subgenus *Arborimus* from *Phenacomys*; renaming *Cleithrionomys* as *Myodes* and *Lagurus* as *Lemmiscus*.
- Order Soricomorpha – updated from the Order Insectivora; removal of the Order Erinaceomorpha and *Erinaceus*
- Order Chiroptera – splitting of the genus *Pipistrellus* into genera *Parastrellus* and *Perimyotis*
- Order Carnivora – merging of the felid genus *Herpailurus* with *Puma*; including *Felis* (bobcat) with *Lynx* while retaining the domestic cat in the genus *Felis*; separation of Family Mephitidae from Mustellidae; separation of *Neovison* from *Mustela*
- Order Artiodactyla – separation of *Bison* from *Bos*

A debt of gratitude is owed to all of those who used the 3rd edition and provided me with comments and corrections that have been incorporated into this edition.  I also owe a very special thank you to my wife Connie Thies for her encouragement and support as I've worked to complete this much needed revision.

# INTRODUCTION TO THE THIRD EDITION

Since its introduction, many people have used this manual as both a reference and tool for working in and teaching mammalogy. However, it has not been available in recent years, having gone out of print. Copies of the second edition are increasingly difficult to locate and those in existence now contain out-of-date information. These facts point to the desirability of a third edition, a possibility suggested on a number of occasions by colleagues and past users of the book. Our intention, with the development of this edition, is to provide a comprehensive treatment of all mammal genera, both native and introduced, occurring in North America north of Mexico, inclusive of marine mammals found in coastal waters.

Primary changes to this edition include a reorganization and update of taxonomy following D. E. Wilson and D. M. Reeder's "Mammal species of the world: A taxonomic and geographic reference" and additional works following its publication in 1993. Several additional exotic genera known to (or potentially) occur in the wild have also been incorporated. As an aid to the user, exotic genera are identified by asterisks in the text and on figure captions. It should also be noted that several of these genera (i.e. *Mus* and *Rattus*) are widespread in distribution and locally abundant whereas others, such as *Erinaceus*, *Meriones*, *Cricetus*, and *Mesocricetus*, may not actually exist in the wild. Old figures have been replaced with line drawings, each including dorsal, ventral, and lateral views of all genera included in the key. As with previous editions, many sources have been drawn upon for both factual and illustrative purposes. Although far from comprehensive, the Selected References section has also been updated to include material used in updating this key, as well as additional books and sources of interest to both the professional and amateur.

Primary changes to the third edition include:
> Order Didelphimorphia - renaming of Order Marsupialia
>
> Order Insectivora - addition of the exotic genus *Erinaceus*; reduction of *Microsorex* to a subgenus of *Sorex*
>
> Order Chiroptera - addition of *Molossus*; *Nyctinomops* split from *Tadarida*; *Idionycteris* split from *Corynorhynus*, which has been renamed from *Plecotus*
>
> Order Carnivora - merging of Orders Carnivora and Pinnipedia into single Order Carnivora; division of felid genera *Felis* and *Lynx* into *Puma*, *Panthera*, *Lynx*, *Herpailurus*, *Leopardus*, and *Felis*; addition of the domestic cat in genus *Felis*; *Thalarctos* reduced to a subgenus of *Ursus*; *Lutra* renamed *Lontra*
>
> Order Cetacea - *Lagenorhynchus* renamed as *Peponocephala*
>
> Order Artiodactyla - addition of *Camelus* and *Lama* in Family Camelidae; change of *Tayassu* to *Pecari*; *Bison* synonomized with *Bos*
>
> Order Rodentia - *Eutamias* reduced to a subgenus of *Tamias*; *Chaetodipus* elevated as separate genus from *Perognathus*; Family Cricetidae combined into Muridae; addition of exotic genera *Meriones*, *Cricetus*, and *Mesocricetus*; *Podomys* elevated to a separate genus from *Peromyscus*
>
> Order Lagomorpha - addition of domestic rabbit, genus *Oryctolagus*; *Brachylagus* elevated to a distinct genus from *Sylvilagus*

Special thanks are due to Kathleen Thies, William Caire, University of Central Oklahoma,

George Baumgardner, Texas A & M University, Tracey Carter, Oklahoma State University, and Joe and Nancy Green, Comanche Spring Ranch, for aid in developing this edition, critical review of the manuscript, and the loan or provision of specimens; Jeremy Jacobs and Diane Nordeck, National Museum of Natural History, for provision of photographic material; and the Sam Houston State University Faculty Enhancement Program for partial funding.

# INTRODUCTION TO THE SECOND EDITION

Since its first appearance in 1951, this manual has been through nine printings, totaling something over 12,000 copies. During this time, a number of exotic species of mammals have been introduced into North America, where they have become established. There have also been some changes in nomenclature in the interim. These facts point to the desirability of a second edition, a possibility that has been suggested on a number of occasions by users of the book.

Enlarging the scope of coverage to include the marine mammals has also been suggested, and it is the tedious production of illustrations and of keys for this new section that has for so long postponed revision. In final form this edition is increased by the addition of the Pinnipedia, Cetacea, and Sirenia. Introduced exotic species of rodents and ungulates have added several genera to those orders, and the keys have been amplified to include subgenera, wherever such exist. The subgeneric keys are not always satisfactory, for the characteristics at this level of classification are often relative rather than absolute. These characteristics may in some cases seem trivial, because better characters may be in some other part of the anatomy than the skull (such as the bushy tail in *Teonoma*). In the Microtinae, the use of subgenera provides a useful solution to the otherwise unsatisfactory alternative of retaining Microtus as a large polytypic genus. The subgeneric key to the ground squirrels obviously comes directly from Howell (1938) with little change. In other cases the sources are more composite.

Special thanks for the gift or loan of specimens, criticism of keys, provision of photographs, and other useful helps, are due to E. Lendell Cockrum, University of Arizona; J. Keever Greer, University of Oklahoma; Robert Hoffman, University of Kansas; James G. Teer, Texas A & M University; William H. Kiel, The King Ranch; Dale Rice and Karl W. Kenyon, U. S. Fish and Wildlife Service; and R. G. Van Gelder and Marie Carrothers, American Museum of Natural History.

# INTRODUCTION TO THE FIRST EDITION

The cranial characters of North American mammalian genera have been described in various publications, but there is no single source wherein the skulls of all forms are diagnosed. Many of the original descriptions are in little-known or inaccessible early publications, and some generic diagnoses (descriptions) are exceedingly difficult to locate. Consequently, anyone having to make extensive identifications must either rely on memory, or have at hand a considerable array of literature.

The author was impressed by the need of some sort of comprehensive treatment of the identification of skulls while undertaking the preparation of keys for use by students of mammalogy. Here, the idea was conceived of preparing a single volume covering all of the genera of the North American continent. Thus far, only the typically terrestrial orders have been included, but it is hoped that a future edition will include the Pinnipedia, Sirenia, and Cetacea as well. The students who have used these keys during the last four years have contributed much by discovering errors and suggesting improvements in the text.

This work should in no sense be considered original, save that here for the first time all of this information is presented in one place. The majority of the genera are represented in the Oklahoma State University Museum of Zoology, and drawings have been made from these, supplemented by references to illustrations and photographs where available. Many such sources have been drawn upon, both for factual and illustrative material. These are listed in the bibliography.

Not all the characters for each genus have been used. In the first place, the author is not an authority on generic characters. In the second place, many of the characters given in other texts are relative, requiring that specimens be compared with those of other genera for the distinction to be appreciated. It is not the purpose of this key to treat all of the distinctions comprehensively, but rather to provide a means for the rapid identification of isolated specimens. For this reason, most of the comparative and relative characters have been omitted. The features that have been included are those that seem to be particularly distinctive, or that provide means for separating forms which are closely similar. As many salient features as possible have been used for each genus in order to facilitate the determination of fragmentary specimens.

Infallibility in a key is a virtually unattainable ideal. It would be most unexpected for this one to prove perfect. Difficulty is most likely to be encountered among the small rodents, in which distinctions are slight and can be fully appreciated only through experience. This is especially true among the Microtinae. In this subfamily, many genera have skulls that are superficially so similar that they may only be distinguished by dental patterns. The genus *Microtus* is especially difficult to characterize without using features shared by other genera. For instance, the subgenus *Pedomys*, using only cranial characters, is virtually indistinguishable from *Pitymys*, and readily so from other species of *Microtus*. It is more closely allied to typical *Microtus* by means of external features. For the purposes of this key, it has proven easier to follow Ellerman in elevating *Pedomys* to full generic rank, and distinguishing it from *Pitymys* by rather trivial means.

In most cases, however, it has been possible to provide characters whereby a genus may clearly be distinguished from other closely related genera. The characters chosen are distinctive, and as a rule show little tendency to intergrade. The use of minor distinctions, as in the Microtinae, *vide supra*, and in the squirrels, is necessitated by the fact that in these instances the salient criterion lies in some other morphological feature rather than in the skull (as in the presence or absence of a baculum in the squirrels).

A number of wild and domestic exotic mammals have been introduced into North America since the advent of the white man. Some of these, such as the fallow deer, chinchilla, European rabbit, nutria and various species of old-world antelope are either purely domestic (caged), or are feral in very restricted localities. Others, among them the cow, horse, pig (including the European wild boar), goat, house rats, and house mouse, are now so widespread in either a semi-domestic or feral state that they are to be met with in many associations. Where this occurs, their skeletal remains may be confused with native forms, and for this reason they have been included in the key. The domestic dog, cat, and sheep are not noteworthy in this respect since they belong to genera represented by native species as well.

It is hoped that this key will prove particularly useful to beginning students in mammalogy, who are usually assigned an array of skulls to be learned; and to wildlife technicians, who may find it a handy reference in the field and laboratory. If this book proves useful to these people it will serve its intended purpose.

The author is especially indebted to the Oklahoma State University Research Foundation for sponsoring this project and providing the time and facilities that have made its completion possible.

The San Diego Natural History Museum, the United States National Museum and Dr. T. C. Carter of Northwestern State College, Alva, Oklahoma, have helped through the loan of specimens used in preparing illustrations. Dr. Randolph L. Peterson of the Royal Ontario Museum, Dr. Robert L. Edwards of Brandeis University, and Dr. R. M. Wetzel of the University of Connecticut all contributed specimens that have been used. Dr. George E. Petrides of Texas A & M College loaned several useful photographs. Dr. David E. Johnson of the U. S. National Museum performed some special investigations on bats of the genus *Choeronycteris*. Dr. George A. Moore of Oklahoma State University critically examined parts of the manuscript. Grateful acknowledgments are due to all of these.

<h1 style="text-align:center">GLOSSARY OF TERMS</h1>

ALISPHENOID - A winglike bone forming part of the lateral wall of the braincase and the posterior wall of the orbit; frequently fused to the basisphenoid.

ALISPHENOID CANAL - A canal on the ventral surface of the alisphenoid bone in front of the auditory bulla.

ALVEOLUS - Socket into which the root of a tooth is set.

ANGLE OF RAMUS (ANGLE OF MANDIBLE) - The posterior projection of the mandible below the mandibular condyle.

ANGULAR PROCESS - A posteroventral projection of the mandible ventral to the coronoid process.

ANTERIOR PALATAL SPINE - The forward-projecting process, formed by the two maxillary bones, that separates the hinder portion of the anterior palatine foramina in rodents.

ANTERIOR PALATINE FORAMINA - See INCISIVE FORAMINA

ANTLER - A deciduous bony head ornament of the frontal bone found in members of the family Cervidae, often only in males.

ANTORBITAL PIT - A pit or depression in the lacrimal bone just in front of the orbit.

ARTICULAR PROCESS - See CORONOID PROCESS

AUDITORY (TYMPANIC) BULLA - The inflated bony capsule that encases the middle and inner ear.


BASAL LENGTH - From anterior border of the median incisive alveoli to mid-ventral border of the foramen magnum.

BASILAR LENGTH - From posterior border of the median incisive alveoli to mid-ventral border of the foramen magnum.

BASIOCCIPITAL - An unpaired bone forming the base of the occipital region.

BASISPHENOID - A median ventral bone lying anterior to the occipital and between the auditory bullae.

BODY - The horizontal portion of the mandible that contains the teeth.

BOSS - A hump; a raised area or protrusion.

BRACHYDONT - Low-crowned; any tooth whose width exceeds the height of the crown above the alveolus.

BRAINCASE - The part of the skull that houses the brain.

BREGMATIC PROCESS - A protrusion of the parietal into the frontal along the suture line on the dorsal surface of the cranium.

BUNODONT - Low-crowned squarish teeth, capped with enamel, and possessing four major cusps arranged in a rectangle.


CANCELLOUS - A term referring to spongy bone; the internal structure may be foamlike, spongelike, or contain numerous large air cells.

CANINE - An enlarged tooth between the incisors and premolars; usually a large stabbing tooth, occasionally bladelike, but sometimes small and similar to the teeth preceding it.

CANINIFORM - Tooth, other than a canine, having the general shape of a canine.

CARNASSIAL - Shearing; in the order Carnivora the term refers to the last upper premolar and the first lower molar which oppose one another like scissor blades and have a shearing action.

CEMENT - Spongy, bonelike material anchoring the tooth root into the alveolus.

CHEEK TEETH - Collectively, the premolars and molars or any teeth located in the skull posterior to the canines.

CONDYLE OF THE RAMUS - See MANDIBULAR CONDYLE.

CONDYLOBASAL LENGTH - From the anterior border of the median incisive alveoli to plane of the posterior border of the occipital condyles.

CORONOID (ARTICULAR) PROCESS - The most dorsal part of the mandible, the part of the vertical ramus dorsal and anterior to the condyle.

CRANIUM - Bones, which collectively make up the upper skull or braincase exclusive of the lower jaw or mandible.

CUSP - A peak or rounded elevation on the crown of a tooth.

CUSPIDATE - Having cusps.


DECIDUOUS - Shed periodically. In mammalian terminology, this term refers to the milk incisors, canines, and premolars that are shed once and replaced by permanent dentition. Also used to describe antlers in the Cervidae (Artiodactyla) which are shed and replaced annually.

DENTAL FORMULA - A convenient way of designating the number and arrangement of mammalian teeth; e.g. I3/3 C1/1 P4/4 M3/3, the full complement for placental mammals. The letters indicate incisors, canines, premolars, and molars. The enumerators give the number of each type in one side of the upper jaw, and the denominators those in one side of the lower jaw.

DENTARY BONE - The lower jaw bone, constituting one half of the lower jaw.

DENTINE - Ivory-like substance beneath the enamel, usually constituting the bulk of the tooth.

DIASTEMA - A vacant space, or gap, between teeth.

EMARGINATE - With a notch or series of notches.

ENAMEL - Exceptionally hard, shiny outer layer on the crown of a tooth.

EROSE - Having the edge irregular or indented, as if eaten away.

ETHMOID VACUITY - An opening in the side of the rostrum formed by the failure of the nasal, lacrimal maxillary, and sometimes the frontal bones to join by means of sutures.

EXTERNAL AUDITORY MEATUS - The external opening into the auditory bulla.

EXTERNAL NARES - The bony external or anterior aperture of the nasal cavity.


FENESTRATE - Having perforations or openings (fenestrae).

FORAMEN - A perforation through a bone for the passage of a nerve or blood vessel.

FORAMEN MAGNUM - The opening in the rear of the skull through which the spinal cord emerges.

FOSSA - A pit, depression, trough, or extensive opening.

FRONTAL - The anteriormost pair of bones covering the brain, situated between the orbits. Actually paired, but often fusing together at maturity.


GLENOID FOSSA - See MANDIBULAR FOSSA.


HAMULUS - A hooklike spine on the posterior corner of the pterygoid bone.

HETERODONT - Having teeth differentiated into various types, i.e. incisors, canines, premolars, and molars.

HOMODONT - Having undifferentiated teeth (see above), usually simple cones.

HORN CORE - The permanent bony spike that serves as a base for a permanent keratinous horn.

HYPSODONT - Teeth with high crowns; usually rootless and ever-growing.


IMPERFORATE - Lacking foramina or fenestrae.

INCISIVE (ANTERIOR PALATINE) FORAMINA - A pair of openings piercing the palate behind the incisor teeth.

INCISORS - Nipping or chiseling teeth at the front of the jaws; upper incisors always root in the premaxillary bone.

INFRAORBITAL CANAL - A passage from the anterior face of the orbit to the side of the rostrum, passing through the base of the maxillary process.

INFRAORBITAL FORAMEN - Same as above, used when the maxillary process is thin and the passage has no length.

INTERNAL NARES - The internal or posterior bony opening(s) of the nasal cavity.

INTERORBITAL REGION - The portion of the cranium lying between the orbits dorsally.

INTERPARIETAL - An unpaired bone at the rear of the cranium, located between the parietals and above and anterior to the occipital.


JUGAL - The bone that forms the mid-section of the zygomatic arch.


KERATINOUS - Composed of a hornlike substance.


LABIAL - On the side next to the lips.

LACRIMAL - A bone, usually small, lying in the anterior face of the orbit and forming part of the orbital rim.

LACRIMAL FORAMEN - A foramen in the lacrimal bone through which the tear duct passes into the nasal cavity, usually near the anterior margin of the orbit.

LAMBDOIDAL (OCCIPITAL) CREST - A transverse bony ridge across the cranium near the posterior border of the parietals.

LAMINATE - With a structure consisting of layers.

LEAST INTERORBITAL BREADTH - The least width between the orbits dorsally.

LINGUAL - On the side next to the tongue.

LOPH - A transverse ridge of enamel across a tooth.

LOPHODONT - Teeth whose crowns have a series of lophs.


MANDIBLE - The entire lower jaw, formed by paired dentary bones.

MANDIBULAR CONDYLE (CONDYLE OF THE RAMUS) - The part at the rear of the mandible that articulates with the upper jaw.

MANDIBULAR (GLENOID) FOSSA - The trough in the squamosal bone for the reception of the mandibular condyle.

MASTOID - The exposed part of the pteromastoid bone, most of which lies within the auditory bulla and encloses the inner ear.

MASTOID BREADTH - The greatest width across the mastoid processes measured at right angles to the long axis of the skull.

MAXILLARY - The bone in the upper jaw that bears the canine, premolar, and molar teeth.

MAXILLARY PLATE - That part of the maxillary bone that forms the flat platelike anterior part of the zygomatic arch; the zygomatic process of the maxillary, especially in rodents.

**MAXILLOVOMERINE NOTCH** - A notch in the posterior border of the vomer, causing a perforation through the bony nasal septum in some rodents; seen by looking into the anterior palatine foramina.

**METALOPH** - The posterior high transverse ridge on the molars of ground squirrels.

**MOLARS** - The posterior teeth in the upper and lower jaws that are non-deciduous.

**NARIAL APERTURE** - Either the anterior or posterior nares.

**NASAL** - The anteriormost pair of middorsal bones forming the roof of the nasal passage.

**NASAL SEPTUM** - A thin, median, vertical partition of bone that divides the nasal cavity into right and left halves.

**OCCIPITAL** - The bone surrounding the foramen magnum and bearing the occipital condyles. Formed from 4 embryonic elements, the basioccipital below, 2 exoccipitals, and the supraoccipital above.

**OCCIPITAL CONDYLE** - A knob on either side of the foramen magnum that articulates with the first vertebra.

**OCCIPITAL CREST** - See LAMBDOIDAL CREST.

**OCCIPUT** - The hinder portion of the skull.

**OCCLUSAL SURFACE** - The crown of a tooth; the grinding surface that faces against the tooth opposing it.

**ORBIT** - The bony socket that contains the eyeball.

**PALATAL BRIDGE** - The solid posterior border of the palate in many Microtinae which seems to bridge over the two troughs or rows of foramina that pass forward from the bridge to the incisive foramina.

**PALATAL LENGTH** - From the anterior border of the median incisive alveoli to the posterior border of the palate (not including the spine).

**PALATAL PIT** - A depression in the lateral border of each palatine bone near the posterior molar.

**PALATAL PROCESS** - A portion of either the premaxillary or the maxillary bone that contributes to the formation of the hard palate.

**PALATAL SPINE** - A median spine projecting posteriorly from the rear border of the hard palate.

**PALATE** - The bony roof of the mouth composed of parts of the premaxillary, maxillary, and palatine bones.

**PALATILAR LENGTH** - From the posterior border of the median incisive alveoli to the posterior border of the palate (not including the spine).

PALATINE - Either of the paired bones that forms the posterior part of the hard palate and walls the anterior part of the interpterygoid fossa.

PALATINE VACUITY - An irregular fenestration or perforation of the palatal portion of the palatine bone.

PARASTYLE - A low, enamel-covered ridge forming the anterior border of the molar teeth in ground squirrels.

PARIETAL - Paired bones roofing the posterior part of the braincase.

PAROCCIPITAL PROCESS - A process projecting ventrally from the lateral border of the occipital bone.  It is lateral to the condyle and behind the auditory bulla.

POSTGLENOID LENGTH - From the plane of the posterior border of the mandibular fossa to the posterior tip of the occipital condyles, measured along the main axis of the skull.

POSTORBITAL BAR - A bony bar between the orbit and the temporal fossa, formed by the union of the superior and inferior postorbital processes.

POSTORBITAL PROCESS - A projection from either the frontal bone (superior) or the jugal bone (inferior), partially separating the orbit from the temporal fossa.

POSTORBITAL WIDTH - Width of cranium immediately behind the superior postorbital processes.

PREMAXILLARY - Paired bones in the front of the upper jaw that bear the incisor teeth.

PREMOLARS - Deciduous teeth posterior to the canines.

PRESPHENOID - An unpaired median bone in the floor of the interpterygoid fossa, with two lateral wings (sometimes called orbitosphenoids) that form part of each orbital wall.

PRISMATIC - With a pattern consisting of sharply-angled triangles, or loops with sharp salient angles.  Geometric in appearance.

PROTOCONE - The major cusp on the labial side of upper molar teeth.

PROTOLOPH - The anterior high transverse ridge (loph) on the molars of ground squirrels.

PTERYGOID - Paired bones, sometimes fused to the basi- and alisphenoids, that form the walls of the posterior part of the interpterygoid fossa.


RAMUS - The vertical portion of the mandible thatramus consists of the mandibular condyle, coronoid process, and angular process.

REENTRANT ANGLE - The indented angles in the sides of hypsodont, prismatic teeth.

ROSTRUM - The portion of the skull anterior to the orbits.

SAGITTAL CREST - A longitudinal median bony crest dorsal to the braincase; often formed by coalescence of temporal ridges.

SALIENT ANGLE - The outward-projecting sharp angles on the sides of hypsodont prismatic teeth.

SECODONT - Having a cutting or shearing action, as in the carnassial pair of carnivores.

SELENODONT - Teeth with longitudinal crescentic ridges of enamel.

SIGMOID - S-shaped.

SPHENOID - Alternate name for the basisphenoid, especially when fused with the alisphenoids and pterygoids.

SQUAMOSAL - A fan-shaped bone on either side of the braincase above the auditory bulla.

SULCATE - Bearing grooves; grooved.

SUPRAORBITAL RIDGE - A beadlike ridge bordering the orbit dorsally.

SUTURE - An immovable line of union between two bones.

SYMPHYSIS - An immovable articulation between the ends of two bones meeting at the midline of the body.


TEMPORAL FOSSA - The large space behind the orbit bounded by the zygomatic arch and the postorbital processes.

TEMPORAL RIDGE - A ridge traversing the top or side of the braincase, marking the dorsal border of the origin of the temporal muscle; may fuse middorsally to form a sagittal crest.

TINE - Any of the spikes or prongs of an antler.

   Brow tine - Rises just above the burr and projecting forward.

   Bez tine - Rises close above the brow tine.

   Trez tine - Rises some distance above the bez tine.

   Royals - Any of the several tines composing the crown or terminal branches.

TRUNCATE - Having the end bluntly squared; appearing to be cut off short.

TUBERCULATE -With rounded elevations or tubercles.

TUBERCULOSECTORIAL - Primitive teeth bearing cusps arranged in asymmetrical triangles, the lower ones having a low posterior heel.

TURBINAL - Any of the several scroll-like ethmoid bones of the nose.

TYMPANIC - A bone that forms the auditory bulla.


UNICUSPID - A tooth bearing only a single cusp.


VOMER - A median unpaired bone in the floor of the nasal cavity.

ZYGOMATIC ARCH - The arch of bone forming the lateral margin of the orbit and the temporal fossa.

ZYGOMATIC BREADTH - The greatest width across the zygomata, measured at right angles to the long axis of the skull.

ZYGOMATIC PLATE - The zygomatic process of the maxillary in the form of a thin flat plate.

ZYGOMATIC PROCESS - A process of either the maxillary or squamosal bone that contributes to the formation of the zygomatic arch.

## KEY TO LABELS ON FIGURES 1 AND 2

**Cranium**

| | | | | |
|---|---|---|---|---|
| 1 | Alisphenoid | | | |
| 2 | Auditory (tympanic) bulla | 20 | Orbit |
| 3 | Basioccipital | 21 | Palatine |
| 4 | Basisphenoid | 22 | Parietal |
| 5 | External auditory meatus | 23 | Paroccipital process |
| 6 | Foramen magnum | 24 | Posterior palatine foramen |
| 7 | Frontal | 25 | Postorbital process |
| 8 | Incisive (anterior palatine) foramen | 26 | Premaxilla |
| 9 | Infraorbital canal | 27 | Presphenoid |
| 10 | Interparietal | 28 | Pterygoid |
| 11 | Jugal | 29 | Sagittal crest |
| 12 | Lacrimal | 30 | Squamosal |
| 13 | Lambdoidal (occipital) crest | 31 | Temporal fossa |
| 14 | Mandibular (glenoid) fossa | 32 | Temporal ridge |
| 15 | Mastoidal process | 33 | Vomer |
| 16 | Maxilla | 34 | Zygomatic arch |
| 17 | Nasal | 35 | Zygomatic plate |
| 18 | Occipital | 36 | Zygomatic process of squamosal |
| 19 | Occipital condyle | 37 | Zygomatic process of maxilla |

**Mandible**

| A | Angular process |
|---|---|
| B | Coronoid process |
| C | Mandibular condyle |
| D | Masseteric fossa |
| E | Body of dentary |

**Dentition**

| i | Incisors |
|---|---|
| c | Canines |
| d | Diastema |
| p | Premolars |
| m | Molars |

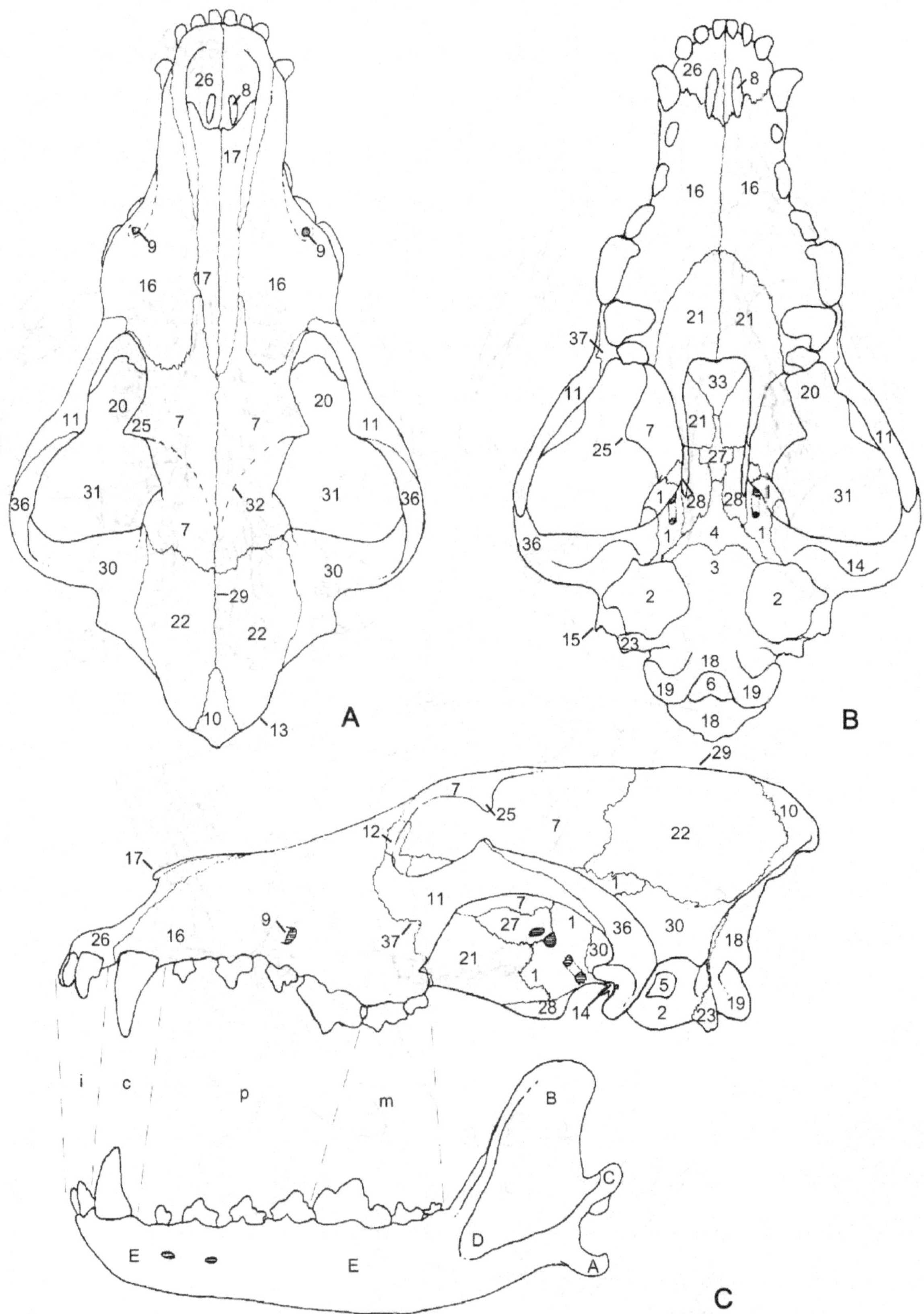



Fig. 1. Cranium and left mandible of *Canis*. A, dorsal view; B, ventral view; C, left lateral view. See page 8 for key to features (modified after DeBlase and Martin, 1981).

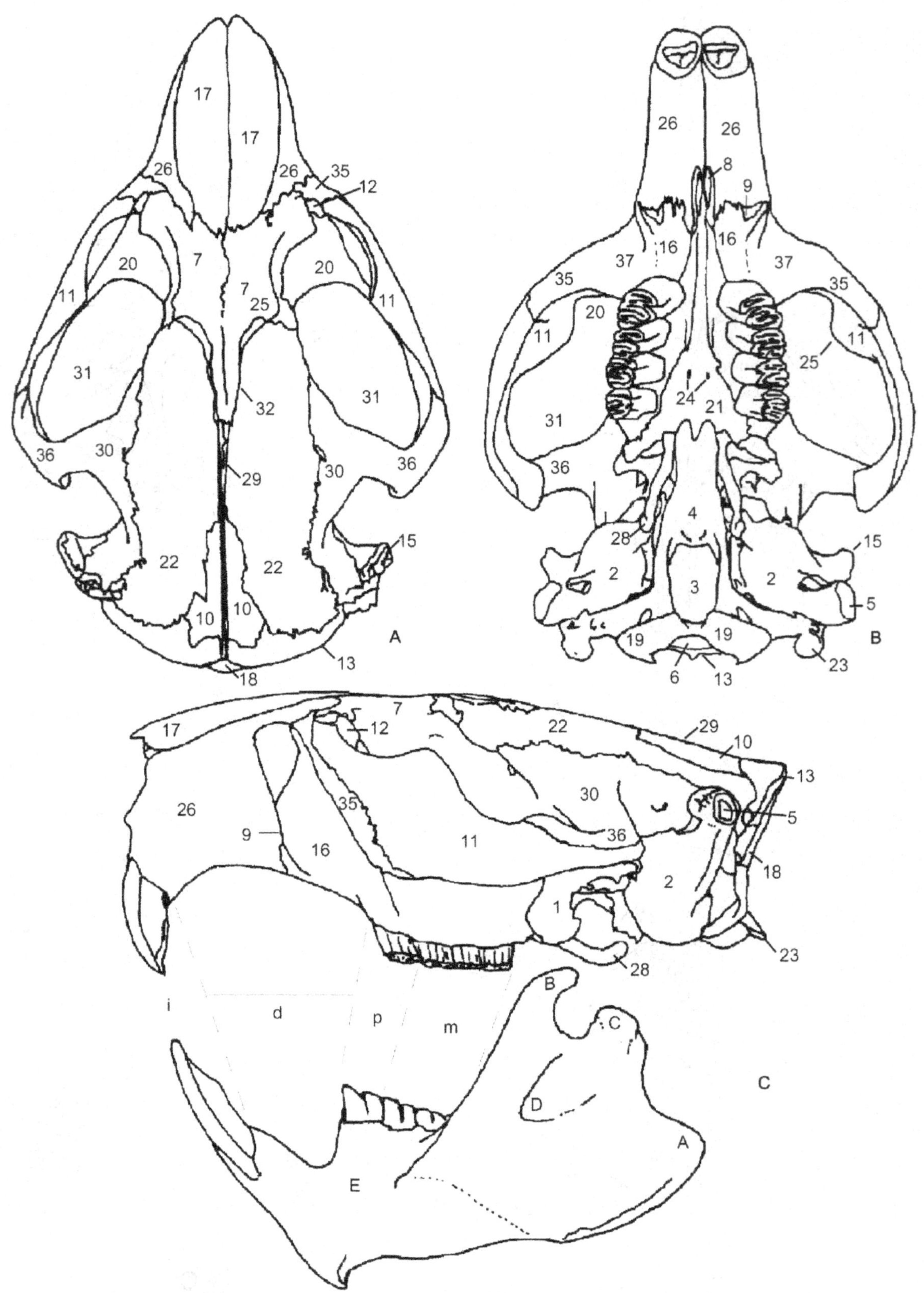



Fig. 2. Cranium and left mandible of *Castor*. A, dorsal view; B, ventral view; C, left lateral view. See page 8 for key to features.

10

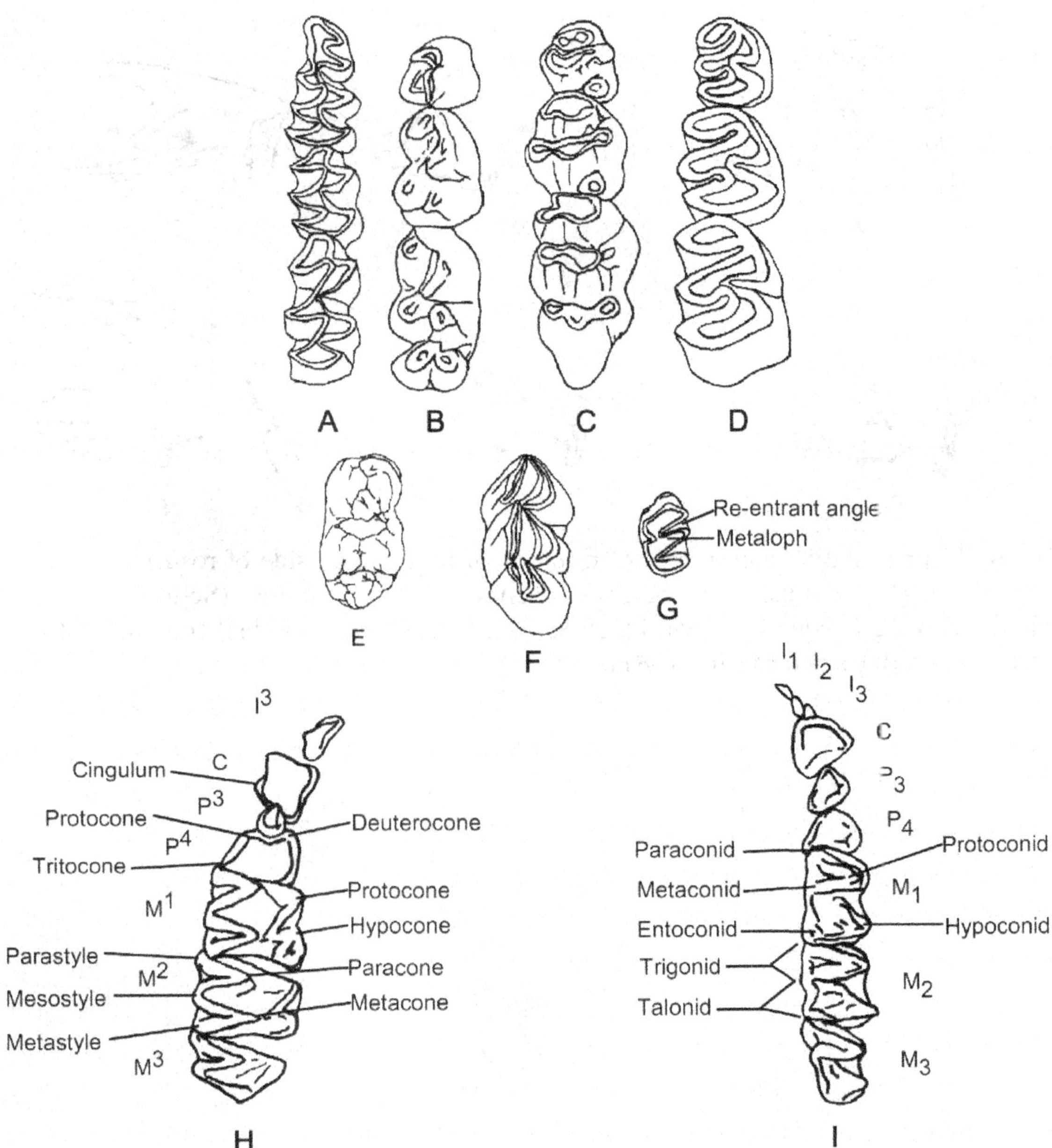



Fig. 3. Representative dental cusp patterns and tooth rows for prismatic cusp pattern (A, *Microtus*); 2 cusp rows (B, *Peromyscus*); 3 cusp rows (C, *Rattus**); laminate pattern (D, *Neotoma*); bunodont crown (E, *Sus**); selenodont crown (F, *Odocoileus*); re-entrant angle and metaloph of first molar (G, *Neotoma*); and upper (H) and lower (I) tooth rows of *Tadarida* with teeth and other features labeled (modified after Fisler, 1970, and Hall, 1981).

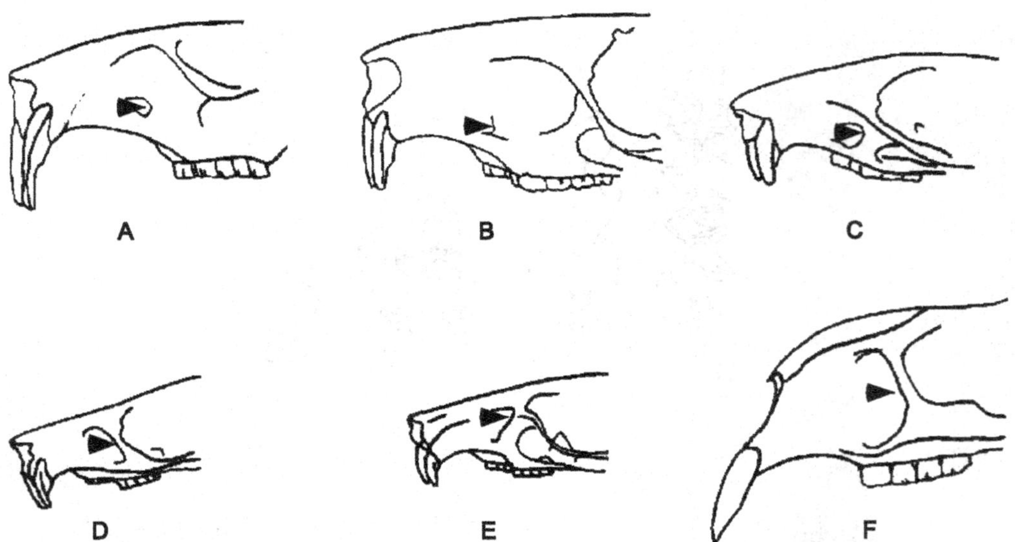

Fig. 4.  Representative appearance of infraorbital foramen on side of rostrum in rodents:  A, *Thomomys* (Geomyidae);  B, *Sciurus* (Sciuridae);  C, *Tamias* (Sciuridae);  D, *Zapus* (Dipodidae); E, *Peromyscus* (Muridae); F, *Erethizon* (Erethizontidae) (modified after Fisler, 1970, and DeBlase and Martin, 1981).

# KEY TO THE ORDERS OF MAMMALS

1. -Teeth absent ................................................................ CETACEA (in part) – page 87
  -Teeth present ..................................................................................................... 2

2. -Teeth heterodont, differentiated into two or more types............................................. 3
  -Teeth homodont, not differentiated into various types ............................................ 12

3. -No conspicuous diastema between the teeth in the front
   of the mouth and the jaw teeth................................................................................ 4
  -A wide diastema present between the teeth in the front
   of the mouth and the jaw teeth................................................................................ 9

4. -Braincase large, rounded; rostrum short; orbits directed
   forward and foramen magnum directed downward ..................... PRIMATES – page 19
  -Braincase small to medium in size; rostrum generally
   elongated; orbits generally directed more laterally and
   foramen magnum directed posteriorly ...................................................................... 5

5. -Canine teeth not larger than any of the teeth preceding
   them.................................................................... SORICOMORPHA – page 45
  -Canine teeth much larger and longer than any of the
   teeth preceding them ............................................................................................... 6

6. -Incisors 5/4, teeth 50 in number; angle of ramus
   inflected inward; palatine vacuities present; part of
   mandibular fossa formed by rear of jugal bone; middle
   and inner ear partially protected by a projection of the
   alisphenoid ....................................................... DIDELPHIMORPHIA – page 16
  -Incisors never more than 3/3, teeth never exceeding 44
   in number; angle of ramus not inflected; palatine bones
   imperforate; mandibular fossa formed entirely by the
   squamosal; middle and inner ears covered by tympanic
   and pteromastoid bones, which usually form an
   auditory bulla ......................................................................................................... 7

7. -Canine teeth triangular in cross-section; lambdoidal and
   sagittal crests prominent, forming a high T-shaped
   ridge; jaw teeth bunodont................................ARTIODACTYLA (in part) – page 75
  -Canine teeth round, oval, or compressed in cross-
   section; occipital crests less prominent and not
   conspicuously T-shaped; jaw teeth various ............................................................... 8

8. -Size small, never exceeding 30mm in length, usually
   much smaller; jaw teeth usually with high-pointed

cusps that are united to form a W-shape; incisors never more than 2 above...................................................CHIROPTERA – page 50

-Size small to large, never less than 27mm, usually much larger; jaw teeth shearing, grasping, or grinding, never with sharp W-shaped cusps; incisors always 3 above, or if 1 or 2, then length greatly exceeding 27mm ......................CARNIVORA – page 59

9.    -Incisors 0/3, upper canines present in some forms; lower canines crowded forward against the incisors and resembling them, and separated widely from the premolars; jaw teeth selenodont; molars emarginate on lingual side; postorbital bar complete.................ARTIODACTYLA (in part) – page 75

-Incisors 1/1, 2/1, or 3/3, canines usually absent; jaw teeth of various sorts, but never selenodont; postorbital bar various...................................................................................... 10

10.    -Incisors 3/3, canines present (may be absent in females, geldings, and hybrids); molars rectangular in outline; postorbital bar complete.................................................PERISSODACTYLA – page 74

-Incisors 1/1 or 2/1; postorbital bar incomplete ............................................................... 11

11.    -Incisors 2/1, jaw teeth 6/5 or 5/5; side of rostrum extensively fenestrate; jugal projecting backward to form a strong projection at posterior end of zygomatic arch ...........................................................................LAGOMORPHA – page 43

-Incisors 1/1, jaw teeth never more than 5/4; side of rostrum not extensively fenestrate; jugal not projecting backward to form a strong projection at posterior end of zygomatic arch .......................................................... RODENTIA – page 20

12.    -Nasal bones strongly telescoped posteriorly; external nares directed vertically upward, with premaxillae and maxillae extending anterior to them in the form of a beak........................................................................................ 13

-Nasal bones in normal position over top of rostrum; external nasal aperture at tip of rostrum and directed forward......................................................................CINGULATA – page 18

13.    -Teeth few to many, usually conical in shape and nearly always extending to the tips of the jaws .......................CETACEA (in part) – page 87

-Teeth quite numerous but rarely more than 6 in either jaw at any one time, squarish in shape, and continuously replacing one another from the rear forward, found in the sides of the jaws only, never at the tips.............................................................................. SIRENIA – page 17

# DIDELPHIMORPHIA

Only one marsupial mammal occurs in North America north of Mexico, the Virginia opossum *Didelphis virginiana* (Didelphidae).  The skull in this genus has many distinctive features.  The dental formula is I5/4 C1/1 P3/3 M4/4 and the molars are tuberculosectorial. The angle of the ramus is directed medially instead of posteriorly as in placental mammals. The nasals are long, and are expanded on the forehead in a diamond shape.  The palatines are perforated in several places to form palatine vacuities.  The jugal extends back far enough to participate in the formation of the mandibular fossa.  The middle ear is not enclosed by the tympanic bone to form an auditory bulla, but is partially enclosed by a cuplike process of the alisphenoid.  The braincase is small and almost tubular, and is surmounted by a high sagittal crest that increases in height and rugosity with age.

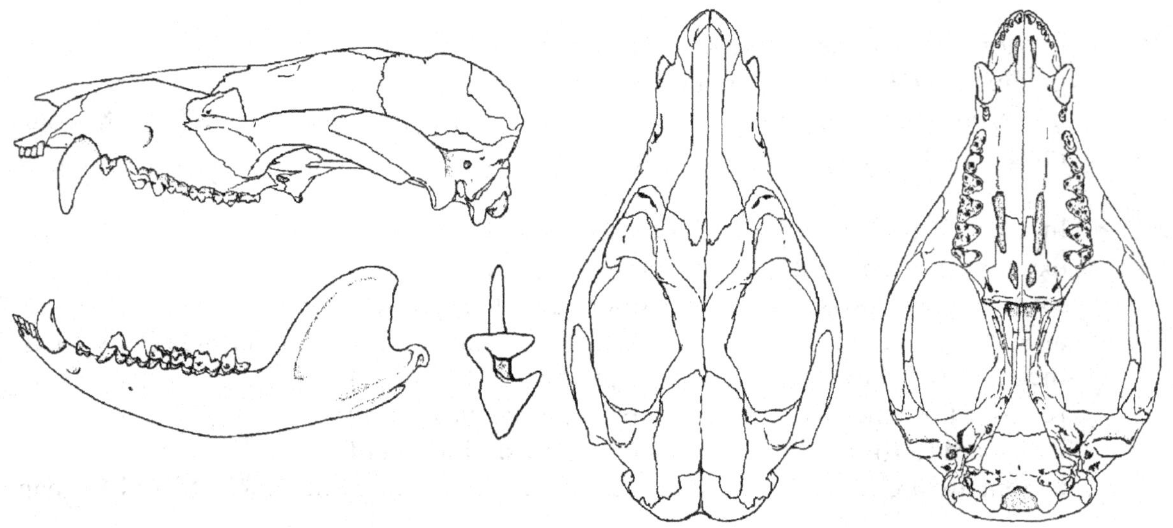

Fig. 5. *Didelphis*
Greatest length of skull 125mm

The manatee is the only representative of this order in North America. *Trichechus manatus* (Trichechidae) may be recognized by the following characters: skull arched, descending steeply from frontals to nasals; zygomata large and massive; jugal greatly developed; orbit small, prominent, almost enclosed by bone; anterior narial aperture very large, lozenge-shaped, and extending posterior to orbits. The mandible is massive, with teeth only in sides and none at the tip, except in young which have an incisive formula of 2/2, the teeth being lost before maturity and not replaced. Molariform teeth in both jaws are similar in character, square, and the enameled crowns are elevated in transverse tuberculated ridges, those in the upper jaw having 2 ridges, those in the lower jaw with 3. Nasals, reduced and receded, are widely separated from one another.

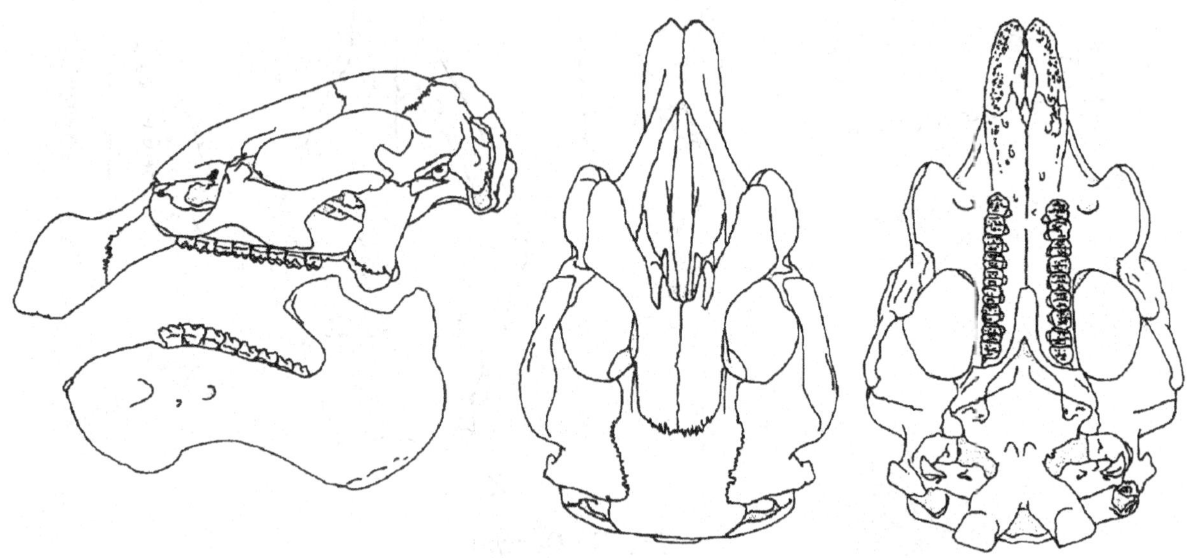

Fig. 6. *Trichechus*
Greatest length of skull 380mm

# CINGULATA

The only North American representative of this order is the nine-banded armadillo, *Dasypus novemcinctus* (Dasypodidae).  The narrow tubular rostrum and the uniform peg-like teeth that have no enamel are most distinctive.  There are usually eight teeth in each jaw, although the number is subject to some variation.  They are limited to the region usually occupied by premolars and molars, none being found in the front of the jaws.  The mandibles are not united by a symphysis.

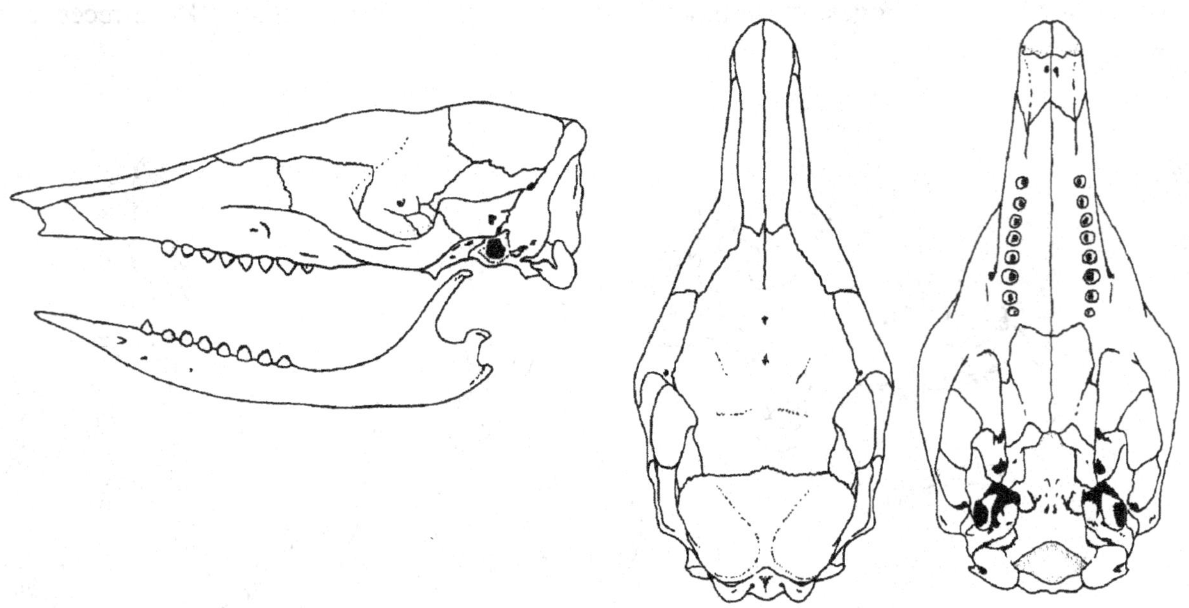

Fig. 7. *Dasypus*
Greatest length of skull 95mm

1.     -Greatest length of skull more than 150mm ................... Hominidae .......... *Homo* (Fig. 8)
    -Greatest length of skull less than 150mm ............ Cercopithecidae ............................ 2

2.     -Facial profile straight; supraorbital arches less prominent .................................................................................. *Chlorocebus* (Fig. 9)
    -Facial profile concave; supraorbital arches more prominent ................................................................................. *Macaca* (Fig. 10)

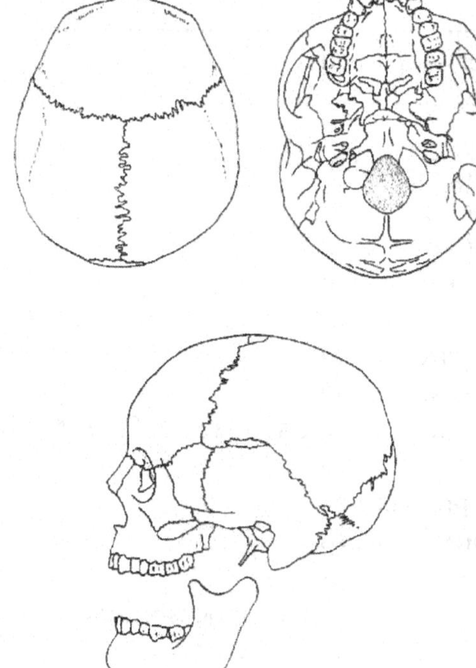

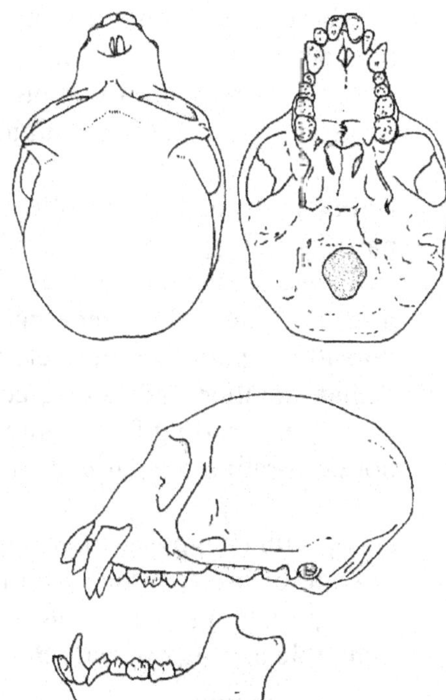

Fig. 8. *Homo**
Greatest length of skull 220mm

Fig. 9. *Chlorocebus**
Greatest length of skull 97mm

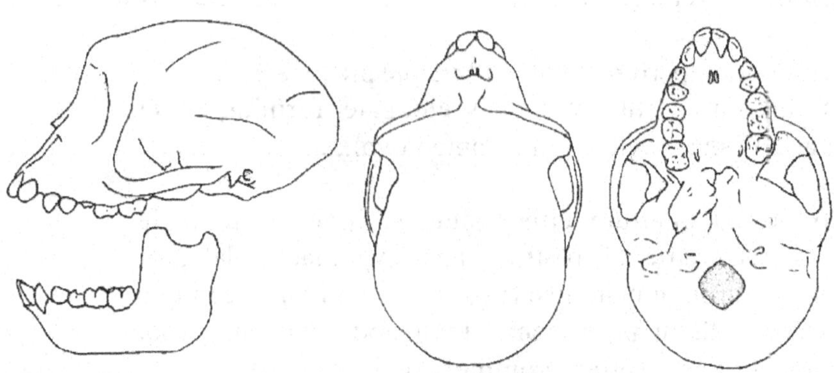

Fig. 10. *Macaca**
Greatest length of skull 105mm

# RODENTIA

1.  -Infraorbital aperture exceeding foramen magnum in
    size; cheek teeth always 4/4; upper cheek teeth tilted
    laterally, lowers tilted medially; tooth rows convergent
    anteriorly ................................................................Suborder Hystricognathi .... 2
    -Infraorbital aperture less than foramen magnum in size;
    cheek teeth 5/4, 4/4, 4/3, or 3/3, usually vertical and set
    in parallel rows (if not, then infraorbital canal small) ......... Suborder Sciurognathi .... 5

2.  -Tooth crowns with a transverse laminate pattern;
    incisors white ...................................................................................................... 3
    -Tooth crowns with a complicated pattern of loops and
    reentrant angles; incisors pigmented................................................................. 4

3.  -Bullae expanded and swollen, nearly meeting ventrally,
    paroccipital processes applied to their posterior
    surfaces; mastoids greatly enlarged and expanded
    dorsally where they penetrate the parietals and appear
    dorsally as paired circular elevations................ Chinchillidae ......*Chinchilla** (Fig. 11)
    -Bullae smaller and separated at midline; paroccipital
    processes standing free from bullae; mastoids small and
    not perforating the parietals dorsally ............................. Caviidae...... *Cavia** (Fig. 12)

4.  -Nasals truncate posteriorly opposite ends of premaxillae
    in front of orbit; paroccipital processes greatly enlarged
    and projecting ventrolaterally; angular process of
    mandible arising opposite $P_4$, forming a broad shelf ............Myocastoridae................
    ............................................................................................ .. *Myocastor** (Fig. 13)
    -Nasals tapered posteriorly, terminating well behind
    premaxillae at midorbit; paroccipital processes small,
    vertical, and terminating above level of bottom of
    bullae; angular process arising at level of $M_2$, not
    forming a broad shelf....................................Erethizontidae ...... *Erethizon* (Fig. 14)

5.  -Infraorbital aperture small and inconspicuous................................................. 6
    -Infraorbital aperture vertically elongate, forming either
    a deep V-shaped notch or a canted oval.......................................................... 33

6.  -Infraorbital aperture either situated in or near angle
    formed by side of rostrum and zygomatic plate, or
    passing through plate itself; base of mandibular condyle
    usually without prominent lateral node enclosing root
    of lower incisor (often prominent in *Sigmodon*)............................................... 7

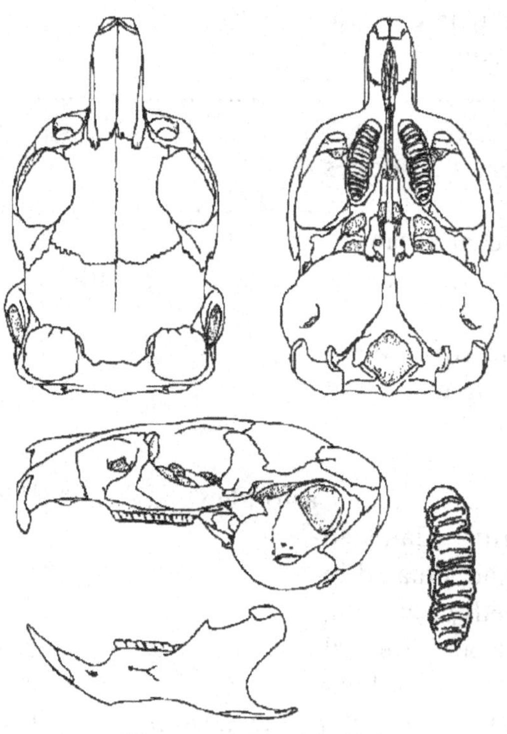

Fig. 11.  *Chinchilla**
Greatest length of skull 50mm

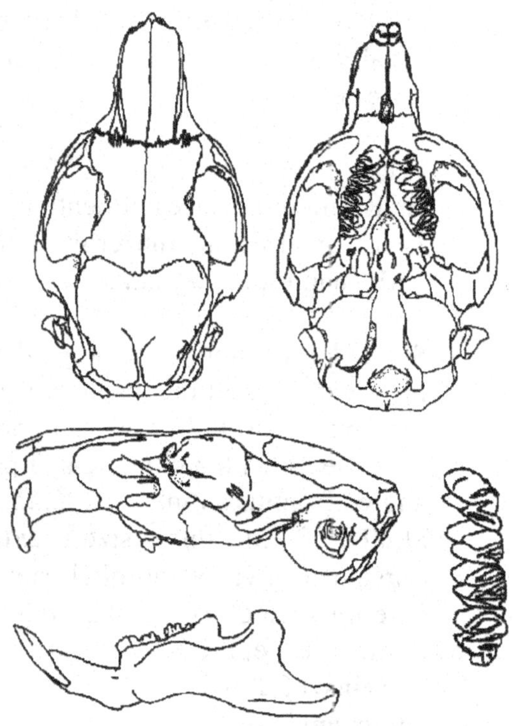

Fig. 12.  *Cavia**
Greatest length of skull 75mm

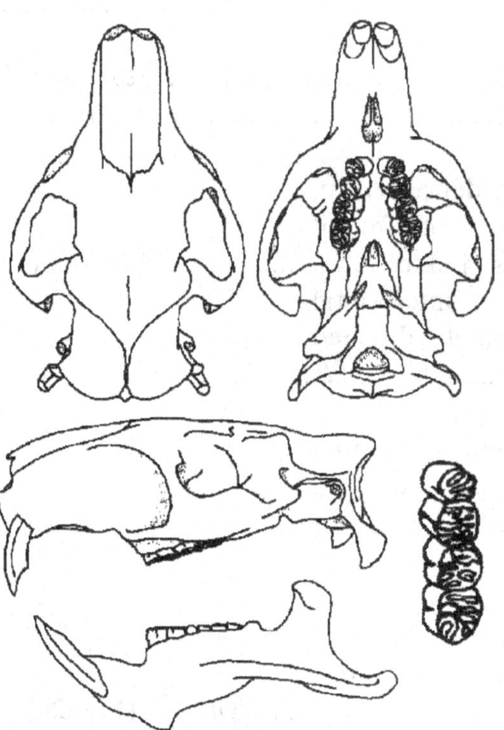

Fig. 13.  *Myocastor**
Greatest length of skull 95mm

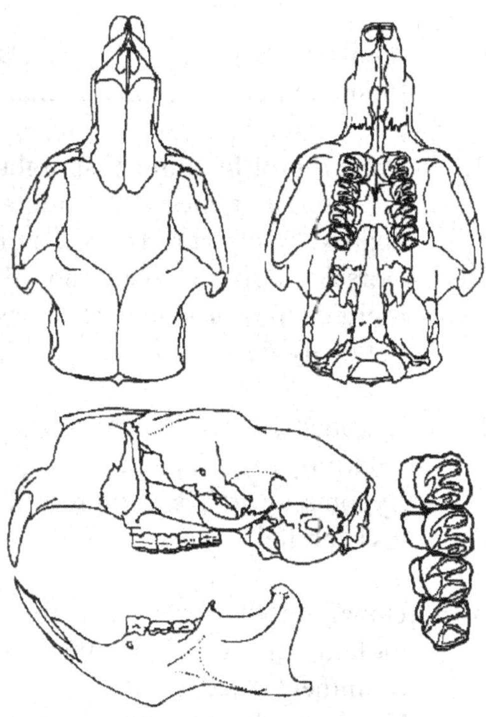

Fig. 14.  *Erethizon*
Greatest length of skull 115mm

-Infraorbital aperture situated on side of rostrum well in front of zygomatic plate; base of mandibular condyle usually with distinct lateral node enclosing root of lower incisor........................................................................................ 26

7.   -Teeth hypsodont, crowns formed of enamel ridges enclosing intraspaces of dentine ..................................................................... 8
-Teeth brachydont, the crowns tuberculate and capped with enamel ...................................................................... Sciuridae......... 9

8.   -Skull large, not extremely flattened dorsoventrally; infraorbital canal inconspicuous and opening on side of rostrum a short distance anterior to zygomatic plate; cheek teeth with a complicated pattern of enamel loops; angle of ramus simple ....................................... Castoridae.......*Castor* (Fig. 15)
-Skull of medium size, extraordinarily flattened dorsoventrally; infraorbital canal round and situated in anterior face of zygomatic arch; cheek teeth circular in outline, except labial margins of upper and lingual margins of lower teeth each bearing a single salient angle; angle of ramus expanded transversely .....Aplodontidae......*Aplodontia* (Fig. 16)

9.   -Infraorbital aperture a foramen piercing zygomatic plate........................*Tamias*....... 10
-Infraorbital aperture a canal passing between zygomatic plate and side of rostrum............................................................................... 11

10.  -Premolars 1/1 ......................................................................(*Tamias*) (Fig. 17)
-Premolars 2/1 ..................................................................(*Eutamias*) (Fig. 18)

11.  -Dorsal profile almost straight, slightly depressed in interorbital region; postorbital processes of frontal approximately at right angles to main axis of skull ...........................*Marmota* (Fig. 19)
-Dorsal profile convex, not depressed in interorbital region; postorbital processes of frontal directed obliquely backward.................................................................................. 12

12.  -Zygomatic arches flattened horizontally and converging anteriorly........................................................................................................ 13
-Zygomatic arches compressed vertically and more or less parallel.................................................................................................... 21

13.  -Upper tooth rows strongly convergent posteriorly; molars much wider than long; lambdoidal crest prominent ...............................................................*Cynomys* (Fig. 20)....... 14
-Upper tooth rows approximately parallel; molars not greatly widened; lambdoidal crest weakly to moderately

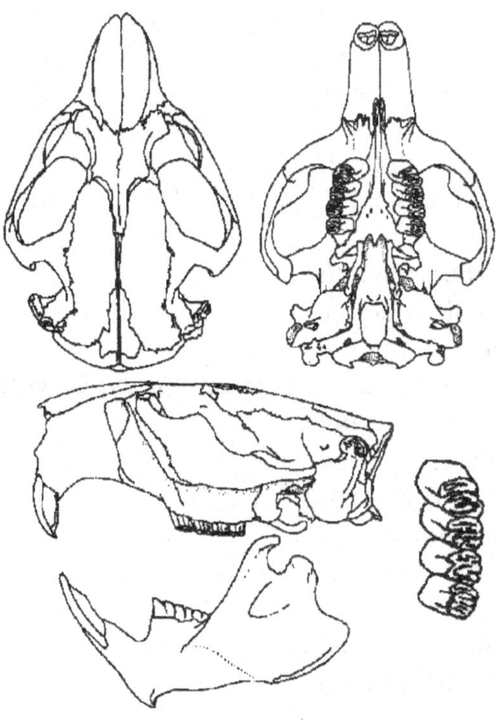

Fig. 15. *Castor*
Greatest length of skull 145mm

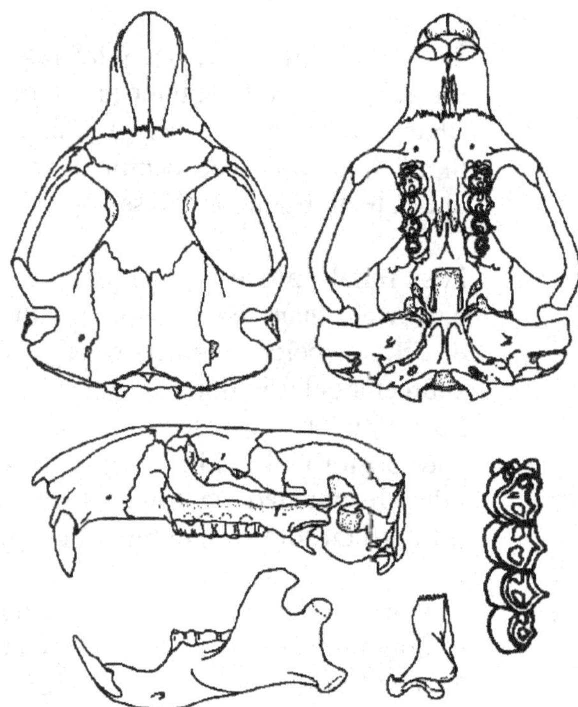

Fig. 16. *Aplodontia*
Greatest length of skull 70mm

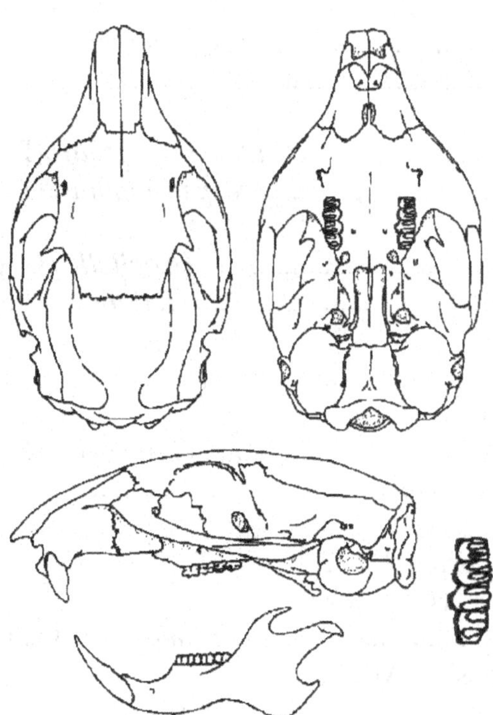

Fig. 17. *Tamias* (*Tamias*)
Greatest length of skull 40mm

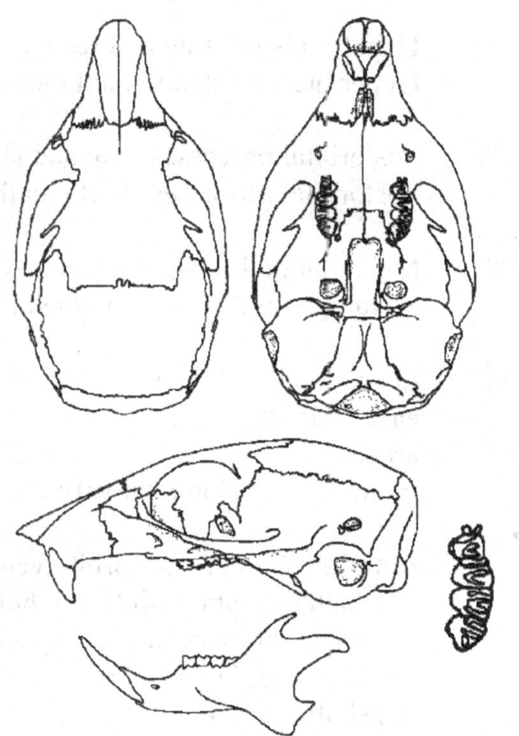

Fig. 18. *Tamias* (*Eutamias*)
Greatest length of skull 30mm

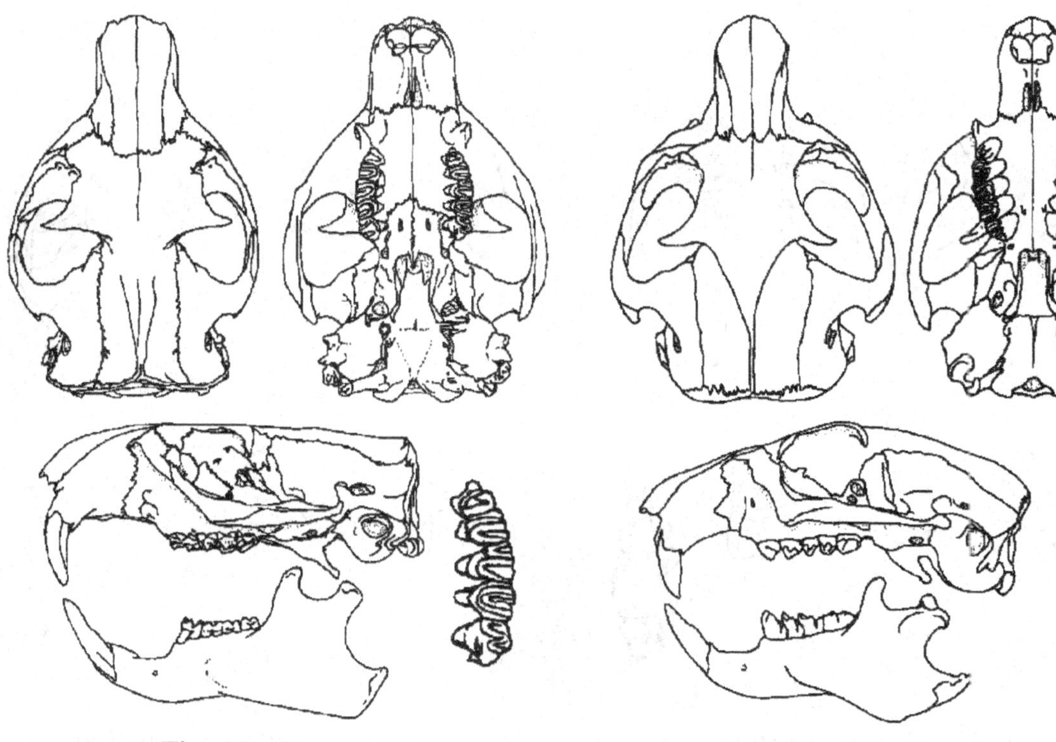

Fig. 19. *Marmota*
Greatest length of skull 90mm

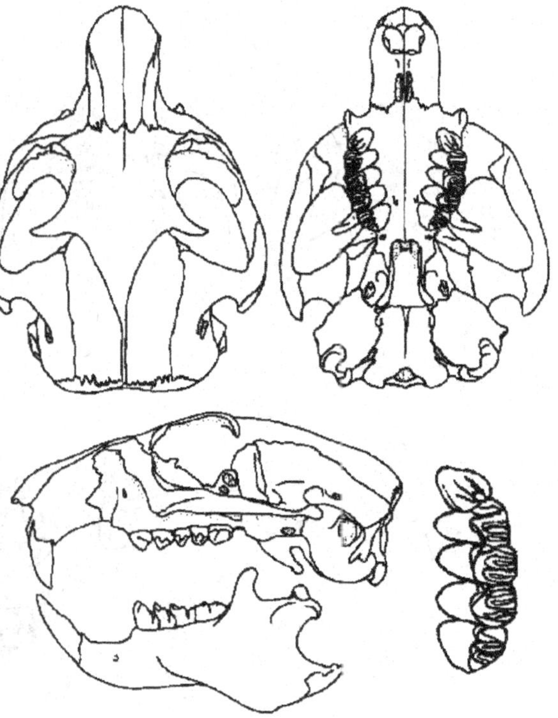

Fig. 20. *Cynomys*
Greatest length of skull 60mm

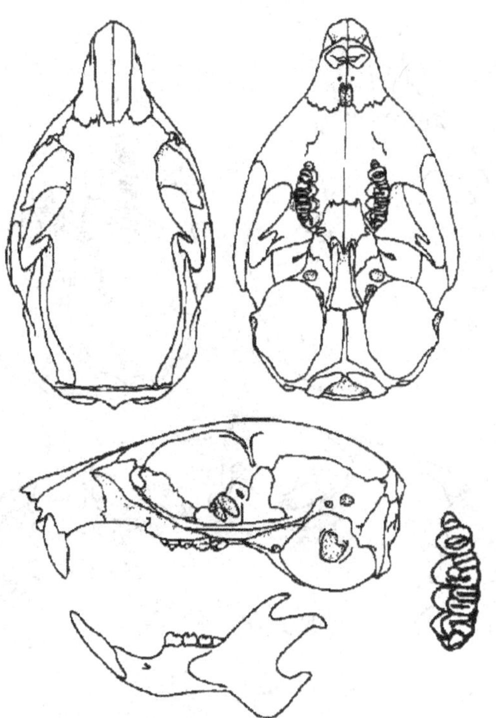

Fig. 21. *Ammospermophilus*
Greatest length of skull 40mm

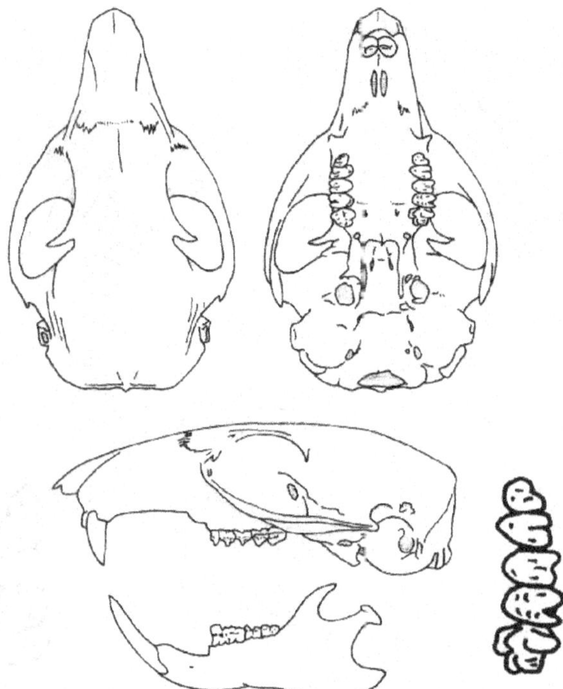

Fig. 22. *Poliocitellus*
Greatest length of skull 55mm

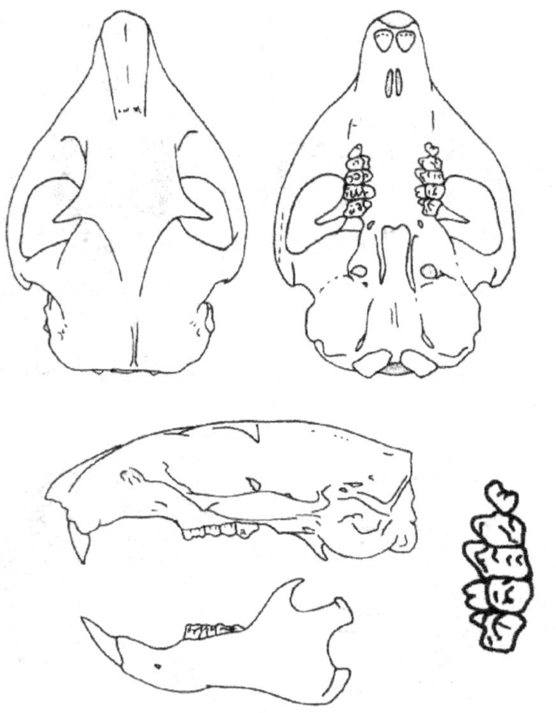

Fig. 23.  *Otospermophilus*
Greatest length of skull 65mm

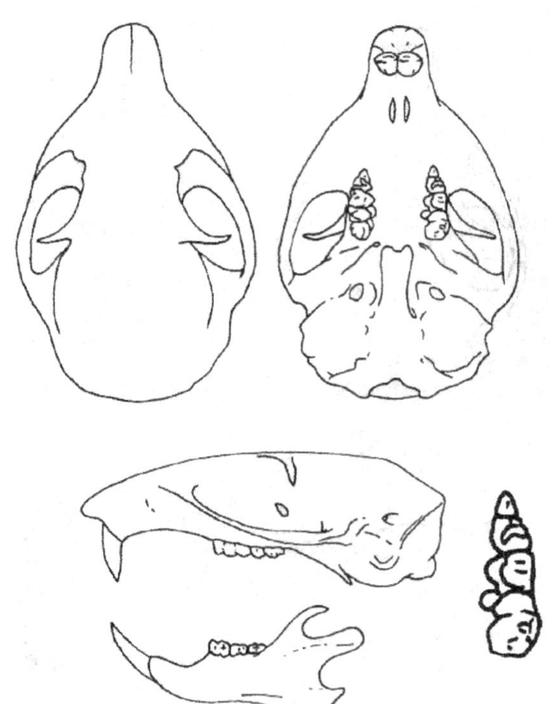

Fig. 24.  *Callospermophilus*
Greatest length of skull 48mm

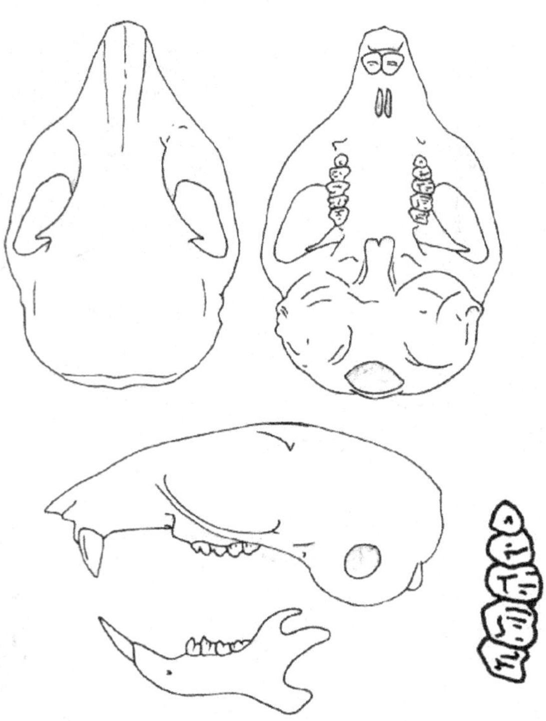

Fig. 25.  *Xerospermophilus*
Greatest length of skull 40mm

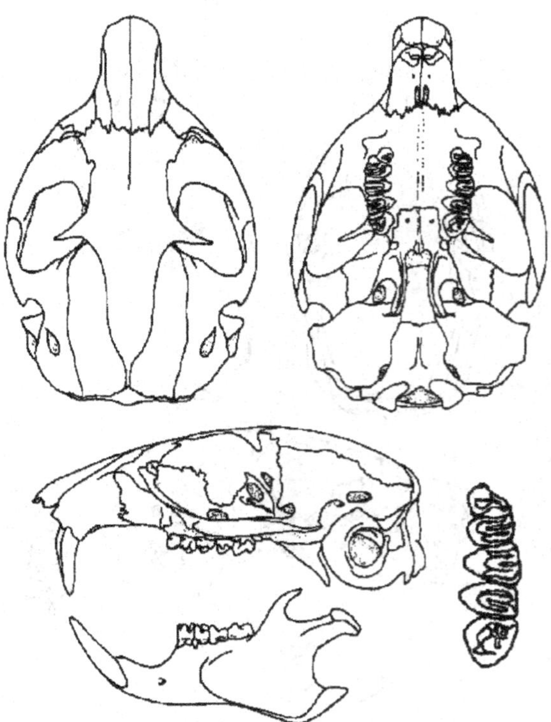

Fig. 26.  *Urocitellus*
Greatest length of skull 65mm

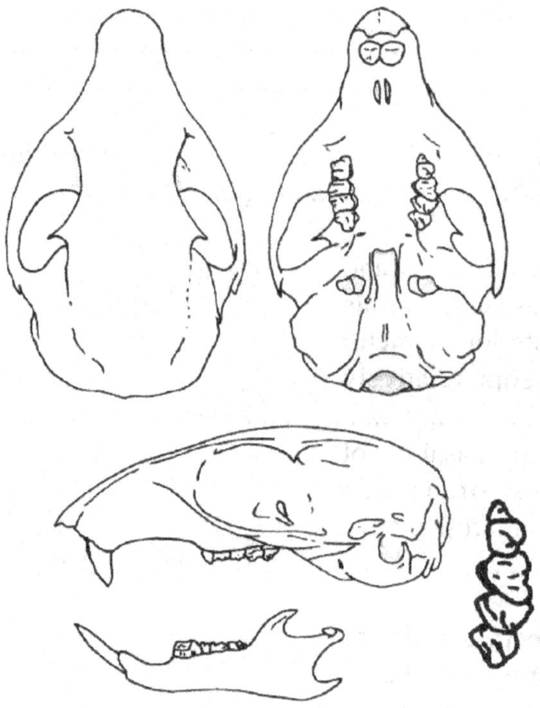

Fig. 27.  *Ictidomys*
Greatest length of skull 52mm

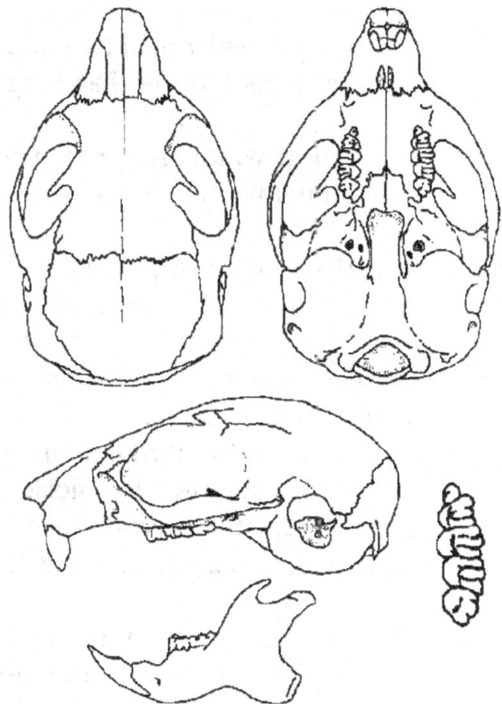

Fig. 28.  *Glaucomys*
Greatest length of skull 34mm

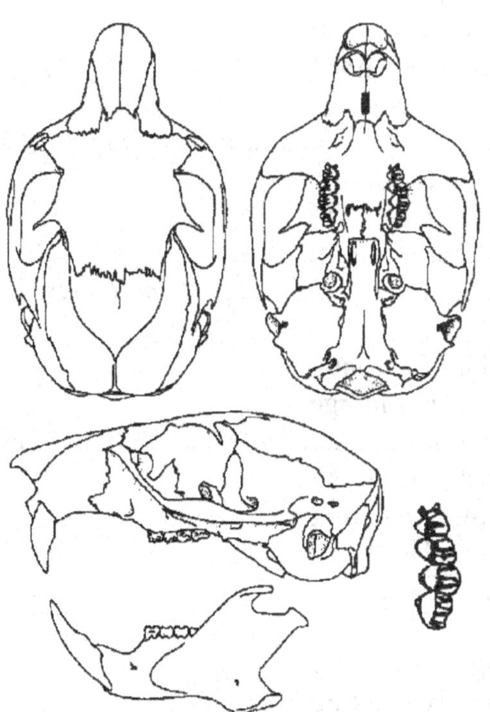

Fig. 29.  *Tamiasciurus*
Greatest length of skull 50mm

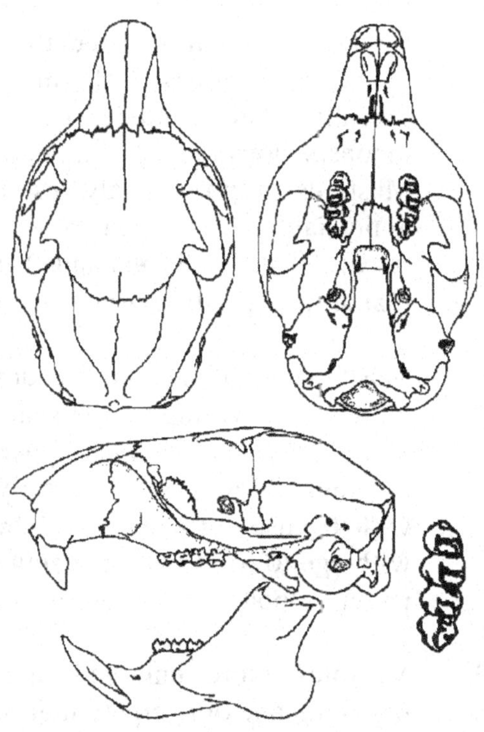

Fig. 30.  *Sciurus*
Greatest length of skull 65mm

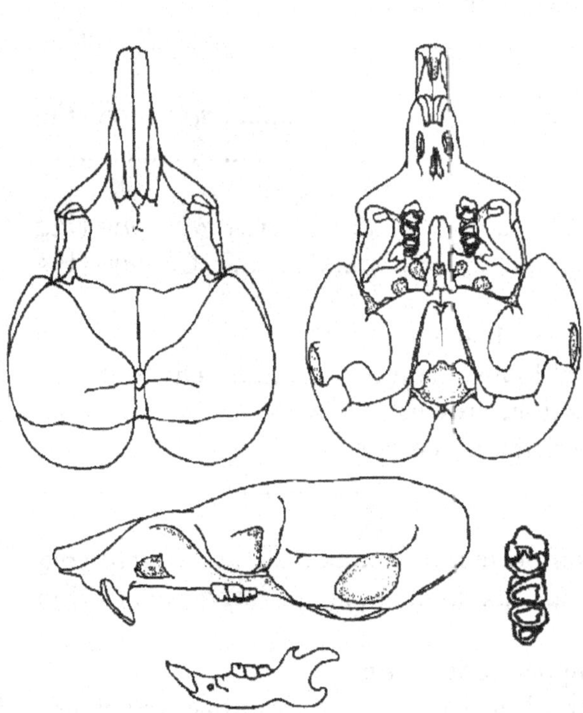

Fig. 31. *Microdipodops*
Greatest length of skull 28mm

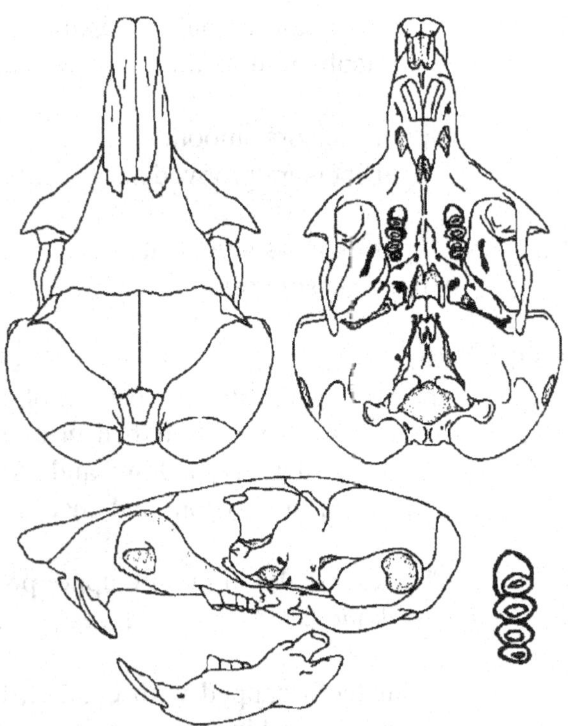

Fig. 32. *Dipodomys*
Greatest length of skull 42mm

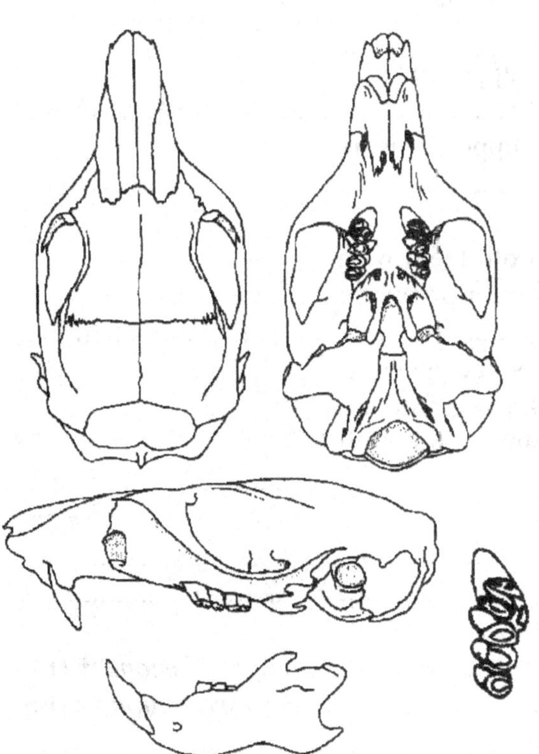

Fig. 33. *Liomys*
Greatest length of skull 35mm

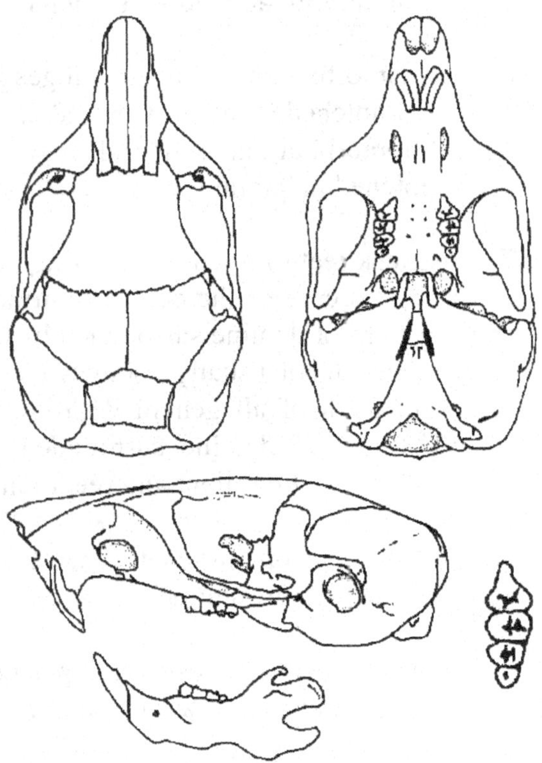

Fig. 34. *Perognathus*
Greatest length of skull 20mm

occiput; auditory bullae separated by nearly full width
of basisphenoid ........................................................ *Chaetodipus* (Fig. 35)

31.   -Upper incisors smooth ........................................ *Thomomys* (Fig. 36)
      -Upper incisors grooved ....................................................................... 32

32.   -Upper incisors unisulcate ................................. *Cratogeomys* (Fig. 37)
      -Upper incisors bisulcate ........................................... *Geomys* (Fig. 38)

33.   -Infraorbital canal oval; crowns of molars brachydont
      and with a complicated pattern of enamel loops .................. Dipodidae ....... 34
      -Infraorbital canal V-shaped in cross-section; crowns of
      molars either hypsodont and with prismatic crown
      patterns or brachydont and cuspidate ................................................... 35

34.   -Cheek teeth 4/3, a minute upper premolar present .................. *Zapus* (Fig. 39)
      -Cheek teeth 3/3 ....................................... *Napaeozapus* (Fig. 40)

35.   -Molar teeth capped with enamel, the upper series with
      cusps arranged in three longitudinal rows (Fig. 151) .............. Muridae (in part) ....... 36
      -Molar teeth either with cusps arranged in two rows or
      with prismatic crown patterns ................................ Cricetidae[1] ....... 37
      [1]Introduced genera identified with an asterisk in
      couplets 39, 46, and 47 belong in Family Muridae.

36.   -Supraorbital and temporal ridges present; upper incisors
      not notched in lateral view ..................................... *Rattus** (Fig. 41)
      -Supraorbital and temporal ridges absent; upper incisors
      notched in lateral view ............................................. *Mus** (Fig. 42)

37.   -Cheek teeth rooted and bearing cusps, in old teeth with
      eroded crowns the occlusal surface without a prismatic
      pattern of dentine surrounded by enamel ............................................ 38
      -Cheek teeth usually rootless (rooted in some genera),
      but teeth of all genera with crowns bearing prismatic
      patterns of dentine surrounded by enamel, and with
      sharply acute salient and reentrant angles ......................................... 49

38.   -Upper incisors grooved ..................................................................... 39
      -Upper incisors smooth ...................................................................... 40

39.   -Zygomatic arches diverging posteriorly ........................ *Meriones** (Fig. 43)
      -Zygomatic arches parallel ............................. *Reithrodontomys* (Fig. 44)

40.   -Molar teeth with flat crowns consisting of cusps
      elongated into transverse lophs, $M^2$, $M^3$, $M_2$, and $M_3$

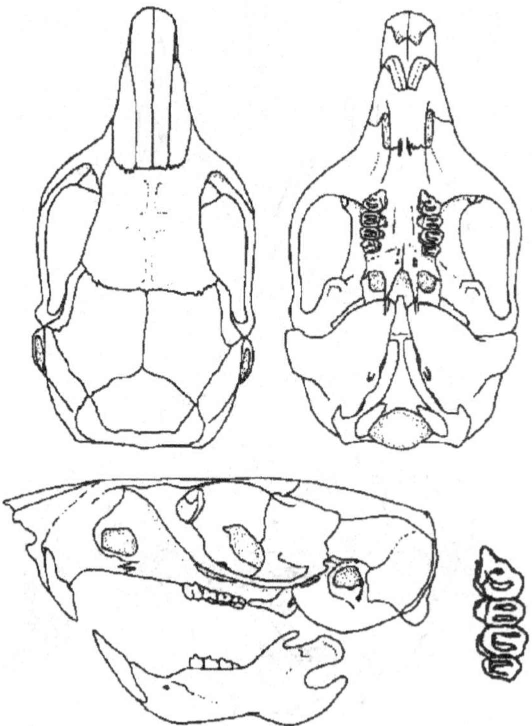

Fig. 35.  *Chaetodipus*
Greatest length of skull 32mm

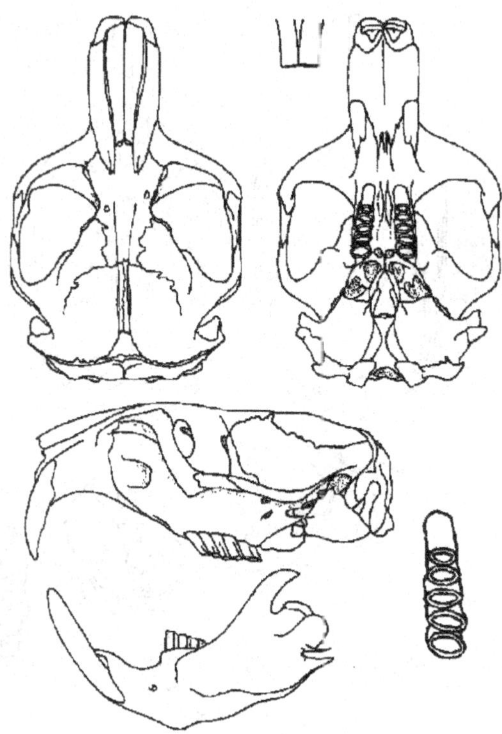

Fig. 36.  *Thomomys*
Greatest length of skull 47mm

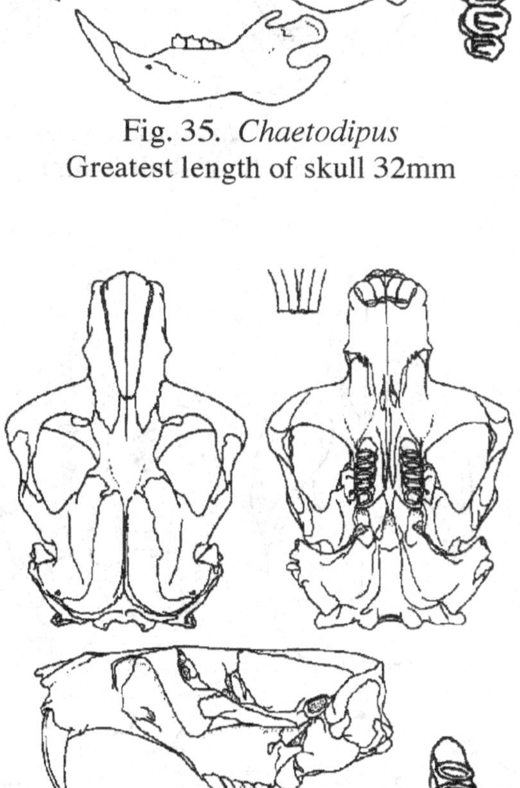

Fig. 37.  *Cratogeomys*
Greatest length of skull 54mm

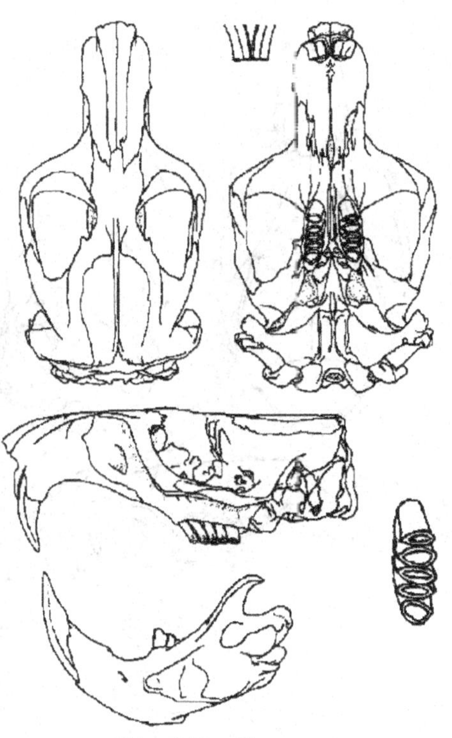

Fig. 38.  *Geomys*
Greatest length of skull 48mm

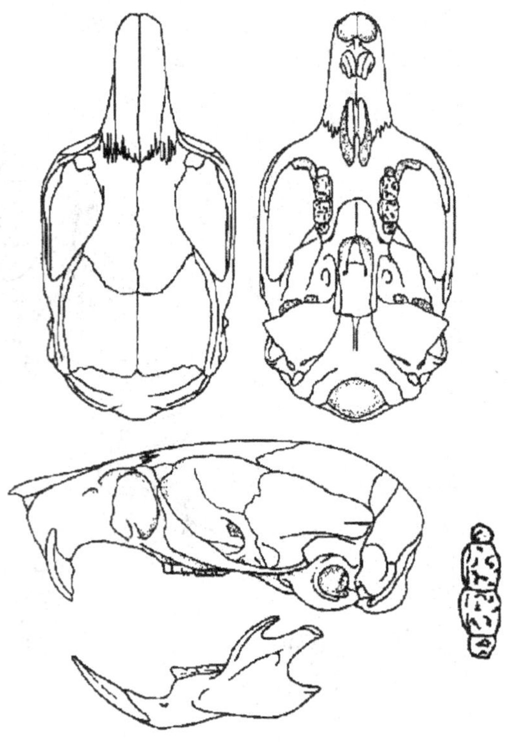

Fig. 39. *Zapus*
Greatest length of skull 24mm

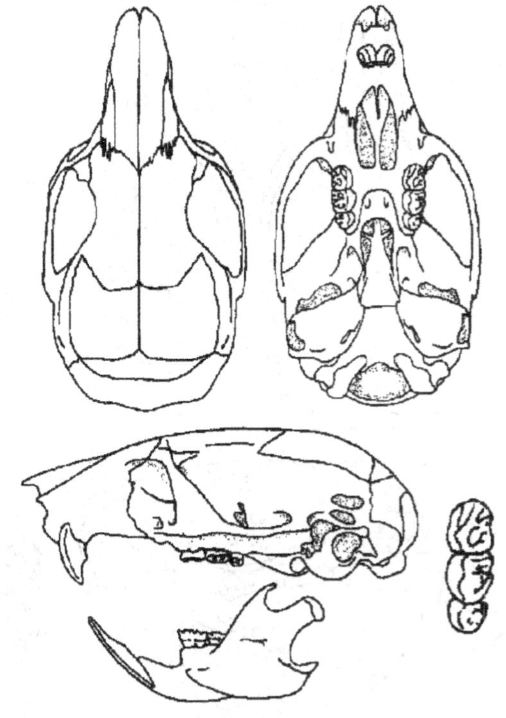

Fig. 40. *Napaeozapus*
Greatest length of skull 24mm

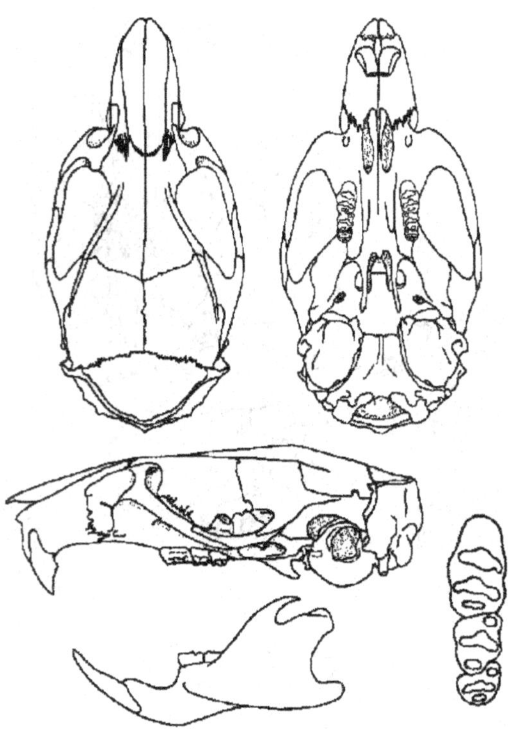

Fig. 41. *Rattus**
Greatest length of skull 45mm

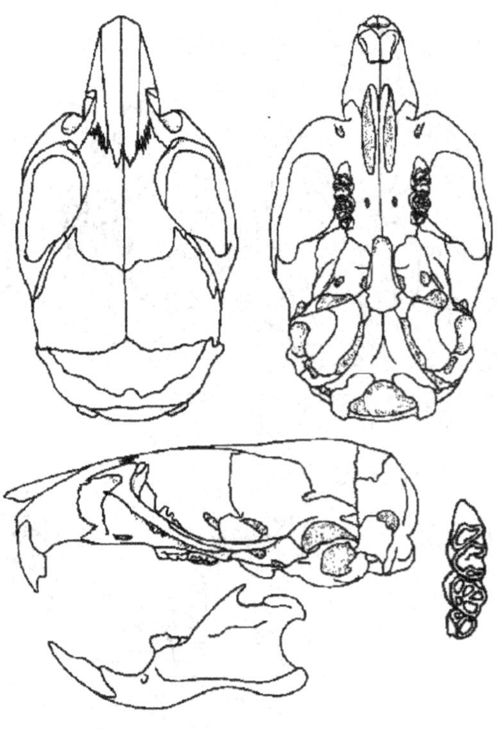

Fig. 42. *Mus**
Greatest length of skull 22mm

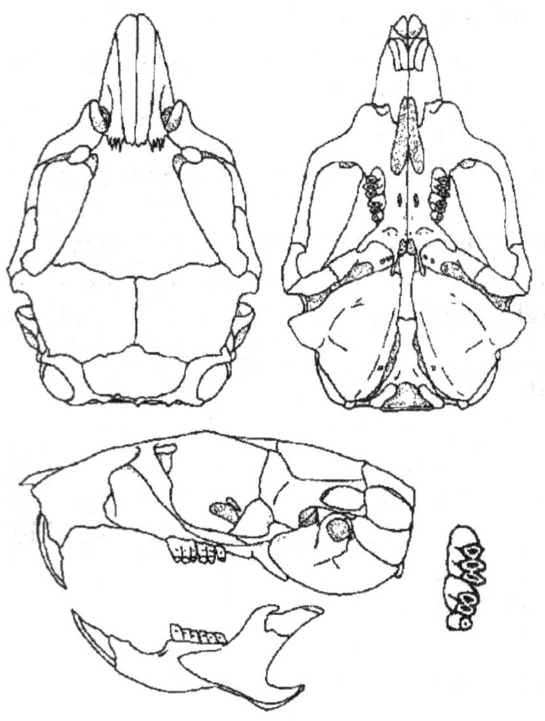

Fig. 43. *Meriones**
Greatest length of skull 35mm

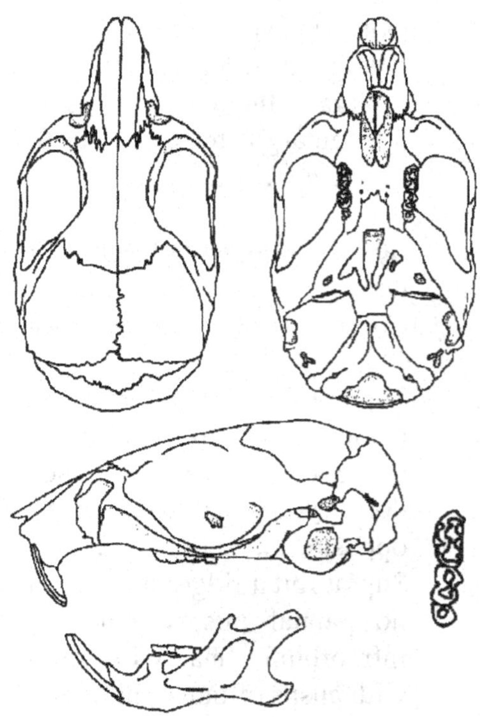

Fig. 44. *Reithrodontomys*
Greatest length of skull 20mm

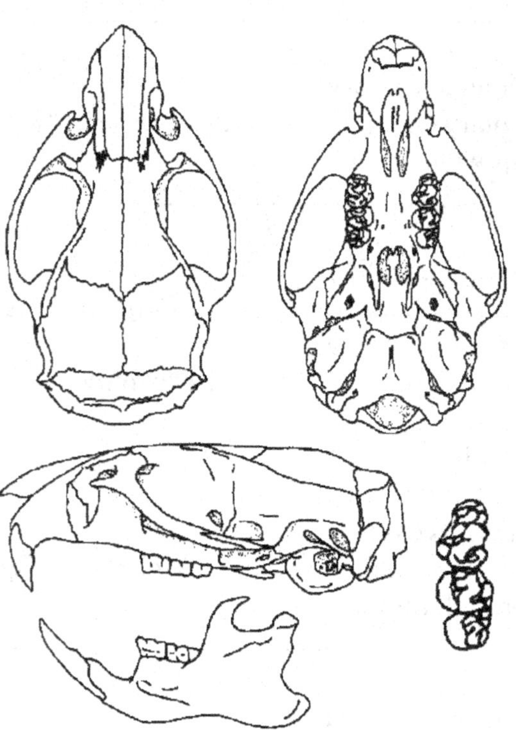

Fig. 45. *Sigmodon*
Greatest length of skull 40mm

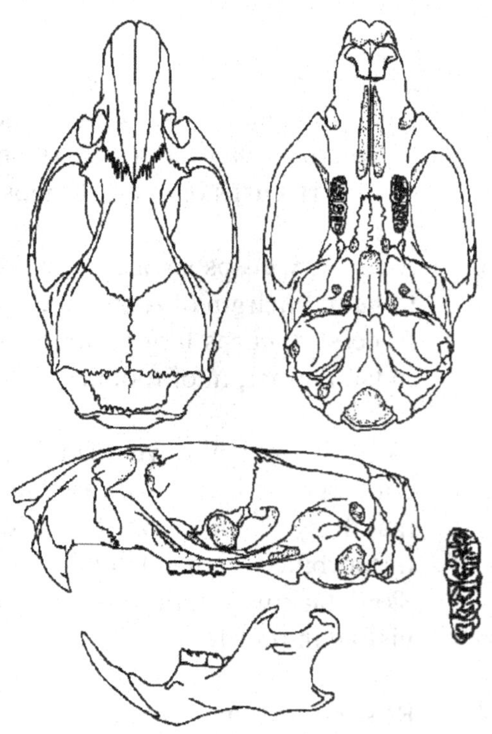

Fig. 46. *Oryzomys*
Greatest length of skull 34mm

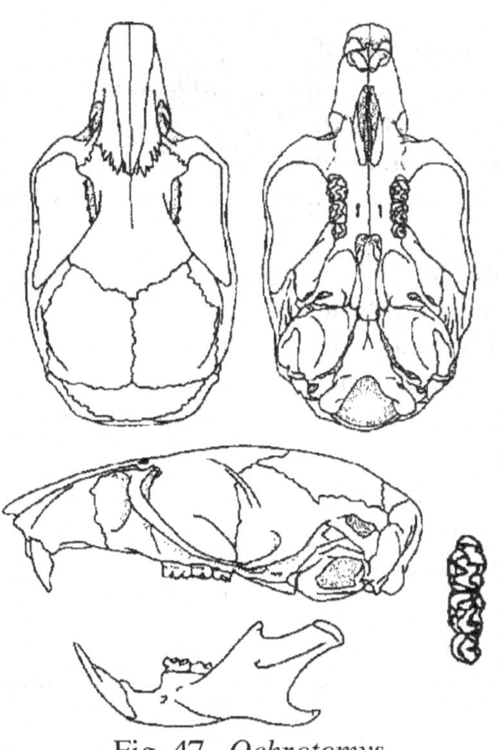

Fig. 47. *Ochrotomys*
Greatest length of skull 25mm

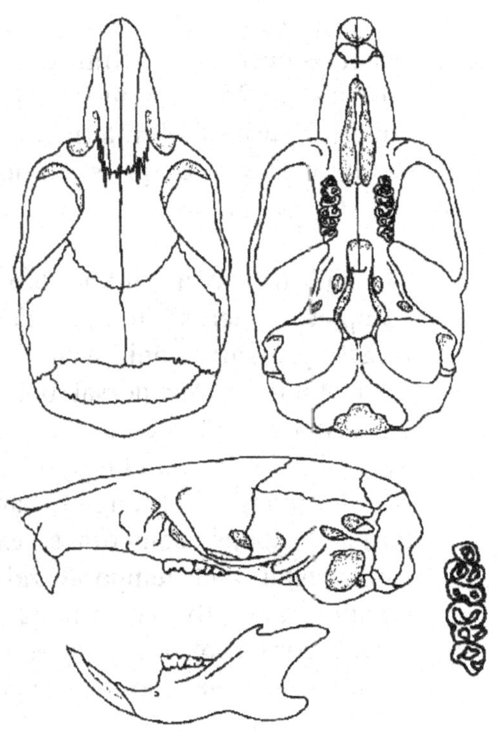

Fig. 48. *Peromyscus*
Greatest length of skull 27mm

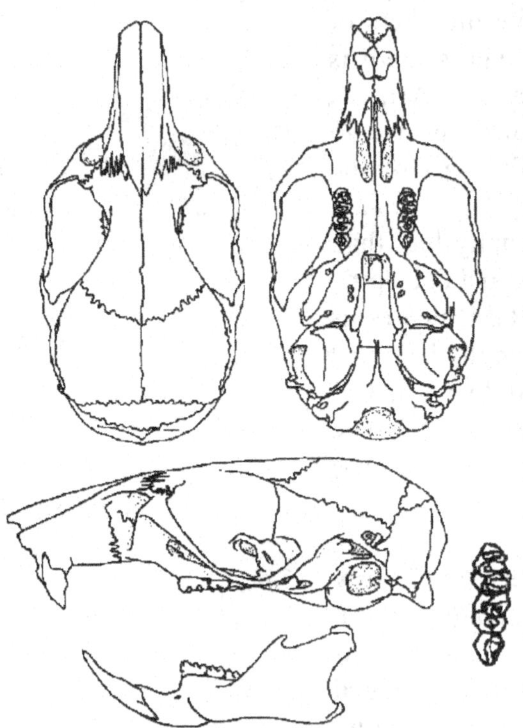

Fig. 49. *Podomys*
Greatest length of skull 30mm

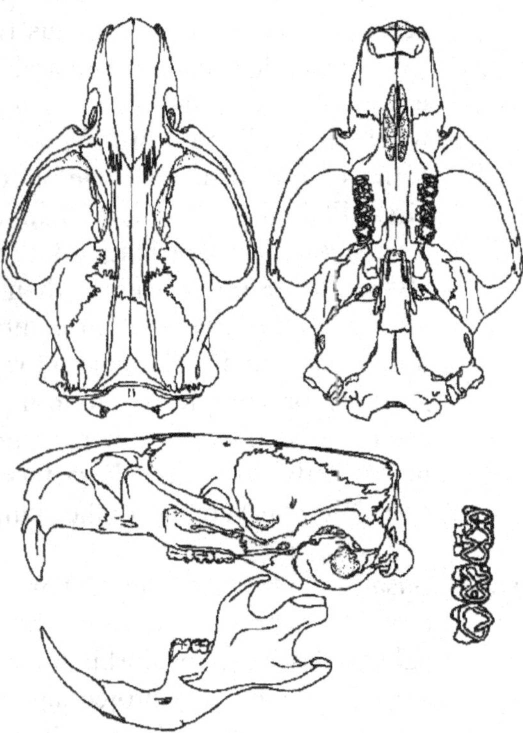

Fig. 50. *Cricetus**
Greatest length of skull 45mm

-Lateral margins of rostrum not parallel......................................................................48

48.    -Anterior cusp of $M^1$ aligned with outer (labial) row of cusps; skull 25mm or more in length.............................................*Onychomys* (Fig. 52)
-Anterior cusp of $M^1$ intermediate in alignment between the two rows of cusps; skull never exceeding 20mm in length..................................................................................*Baiomys* (Fig. 53)

49.    -$M^1$ and $M^2$ with middle loop undivided, extending completely across tooth; supraorbital and temporal ridges present; cranium not constricted abruptly in interorbital region; dorsal border of infraorbital canal emarginate; teeth rooted..................................*Neotoma* (Fig. 54).......50
-$M^1$ and $M^2$ with middle portion divided into a number of sharp-angled triangles, and usually with only terminal loops sometimes extending across tooth; supraorbital and temporal ridges absent or present; cranium abruptly constricted in interorbital region; dorsal border of infraorbital canal virtually straight; molars sometimes rooted, sometimes rootless......................................................................................51

50.    -Size large, basilar length 38-46mm, usually more than 43mm; anterior palatal spine pointed; maxillovomerine notch present, nasal septum incomplete (Note: The characteristics of the palatal spine and nasal septum serve to separate the subgenus from some members of (*Neotoma*) that are of comparable size. The system is unreliable for some areas, e.g. northeastern Arizona where a small race of (*Teonoma*) is equal in size to some of the local (*Neotoma*); found only west of the Great Plains)...............................................................*(Teonoma)*
-Size smaller, basilar length 33-38mm, usually less than 38mm; if 38mm or more, then anterior palatal spine bifurcate and nasal septum complete, OR $M^3$ nearly as broad as $M^1$ and with the inner reentrant angle deep and partially or completely dividing the middle loop, OR found only in northeastern United States (Note: The more easily discerned characters separating (*Teonoma*) and (*Neotoma*) are not in the skull)......................*(Neotoma)*

51.    -Basal length exceeding 50mm; maxillary tooth row more than 14mm; posterior border of palate without palatal bridges; postorbital processes nearly right-angled in shape and projecting into orbit as thin-edged shelves (as in *Neofiber*); bullae without internal bony trabeculae; molars rooted..................................*Ondatra* (Fig. 55)

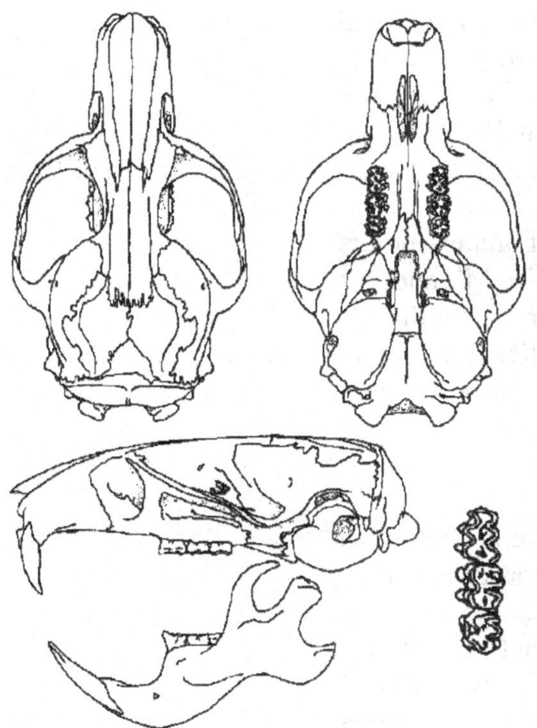

Fig. 51. *Mesocricetus**
Greatest length of skull 35mm

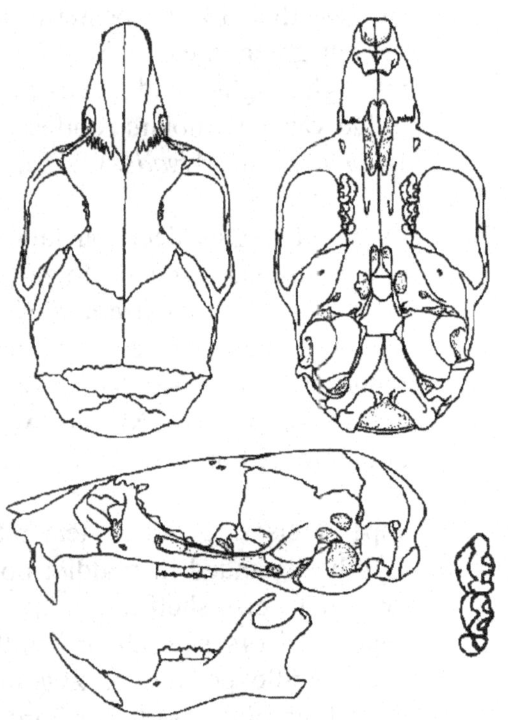

Fig. 52. *Onychomys*
Greatest length of skull 28mm

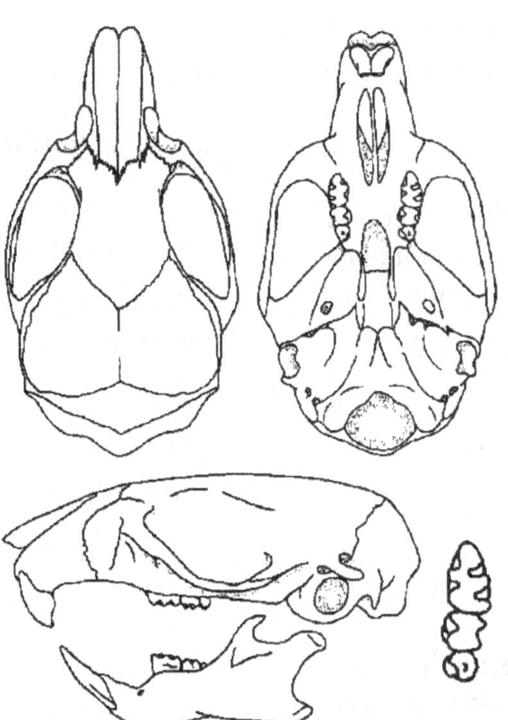

Fig. 53. *Baiomys*
Greatest length of skull 19mm

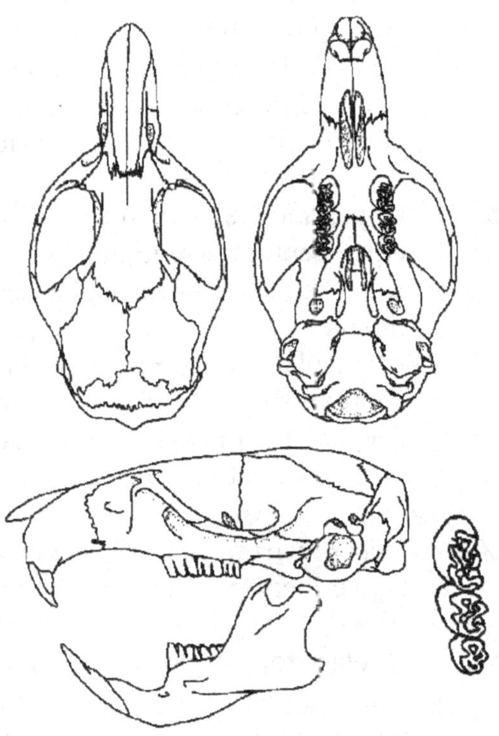

Fig. 54. *Neotoma*
Greatest length of skull 50mm

37

-Basal length less than 45mm, usually much less; tooth row less than 13mm; posterior border of palate bridged (except sometimes in *Neofiber*); postorbital processes not right-angled and shelf-like (except in *Neofiber*); bullae various; molars rootless (except in *Phenacomys, Arborimus*, and *Myodes*) ........................................................ 52

52.   -Reentrant angles deep on labial side of upper molars and on lingual side of lower molars, in $M^1$ and $M^2$ cutting across to extreme inner border of tooth; $M^3$ formed of 4 transverse loops; molars rootless ..................... 53
-Reentrant angles not deeper on one side in both upper and lower molars; $M^3$ not with 4 transverse loops; molars rooted or rootless.............................................................. 55

53.   -Upper incisors grooved near lateral margin; zygomata slightly expanded in middle; posterior border of palate not a transverse shelf............................................................... 54
-Upper incisors smooth or faintly grooved on anterior face; middle portion of zygomata vertically enlarged into a high plate; posterior border of palate a transverse shelf with a slight median spine........................... *Lemmus* (Fig. 56)

54.   -Mandibular molars with closed triangles on buccal side, buccal reentrant angles well developed; median palatal spine poorly developed; upper incisors truncate-edged............... *Synaptomys* (Fig. 57)
-Mandibular molars without buccal triangles, and buccal reentrant angles obsolete; strong median palatal spine; cutting edge of upper incisors notched ...................................... *Mictomys* (Fig. 58)

55.   -Upper molars with reentrant angles on both sides of equal depth, lower molars with angles deeper on lingual side; molars rooted; bullae without bony trabeculae internally ................................................................... 56
-All molars with reentrant angles on both sides of approximately equal depth, lower molars with angles deeper on lingual side; molars rootless except in (*Myodes*); bullae various.............................................. 57

56.   -Upper incisors pointing downward ............................... *Phenacomys* (Fig. 59)
-Upper incisors recurved and pointing posteriorly ......................... *Arborimus* (Fig. 60)

57.   -Molars rooted; posterior border of palate a thin-edged straight shelf, continuous from side to side between posterior molars; bullae without internal bony trabeculae .................................................................... *Myodes* (Fig. 61)

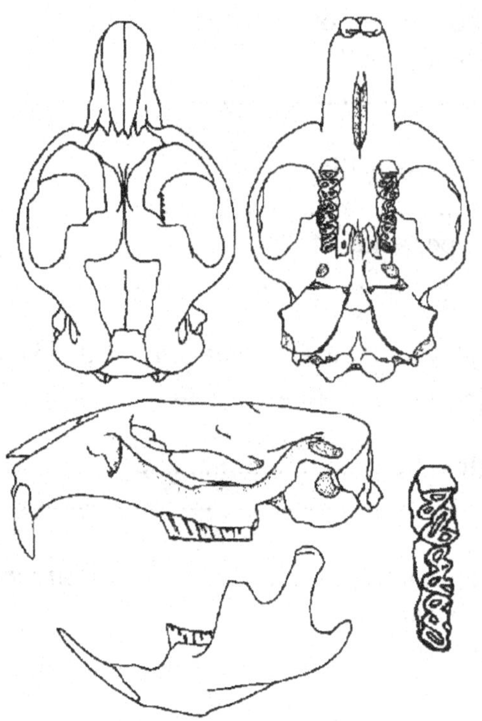

Fig. 55. *Ondatra*
Greatest length of skull 65mm

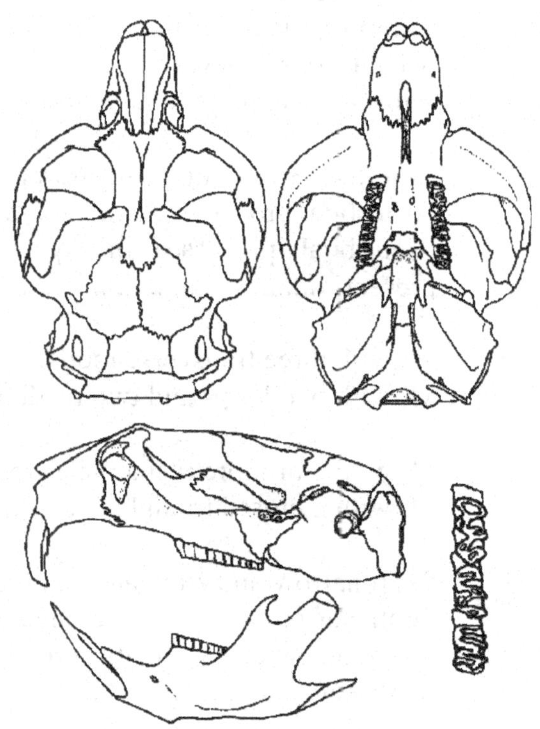

Fig. 56. *Lemmus*
Greatest length of skull 30mm

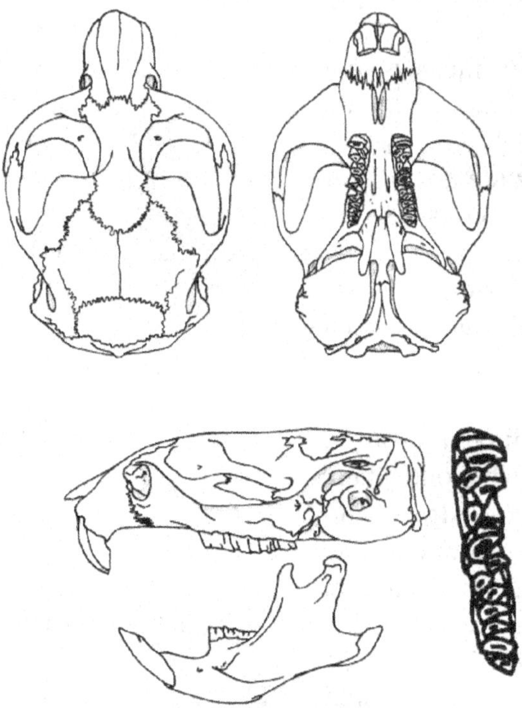

Fig. 57. *Synaptomys*
Greatest length of skull 30mm

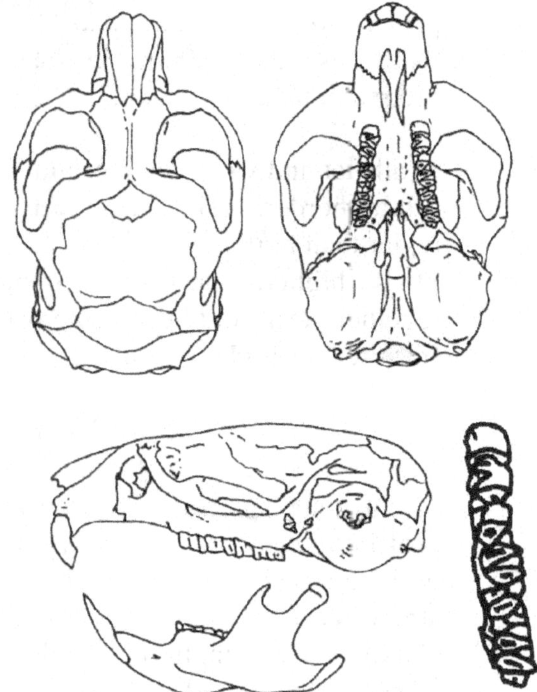

Fig. 58. *Mictomys*
Greatest length of skull 26mm

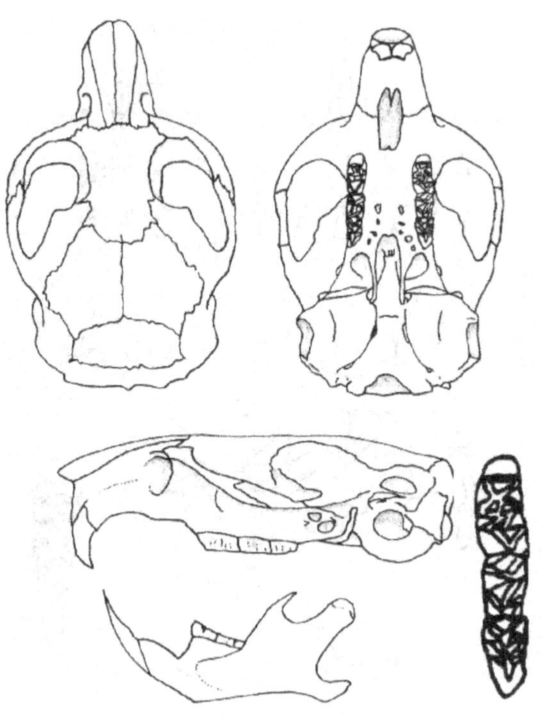

Fig. 59. *Phenacomys*
Greatest length of skull 27mm

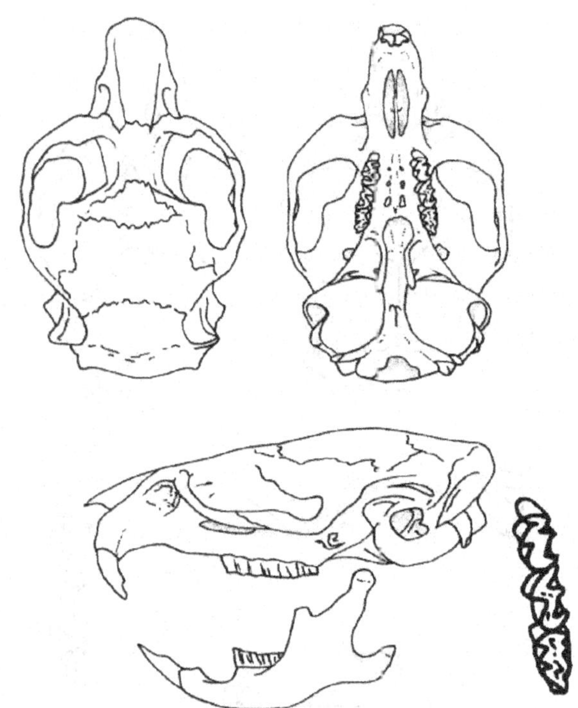

Fig. 60. *Arborimus*
Greatest length of skull 27mm

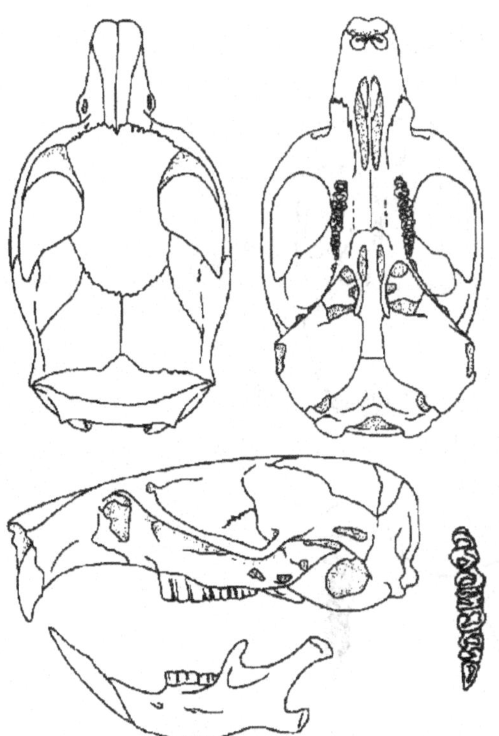

Fig. 61. *Myodes*
Greatest length of skull 25mm

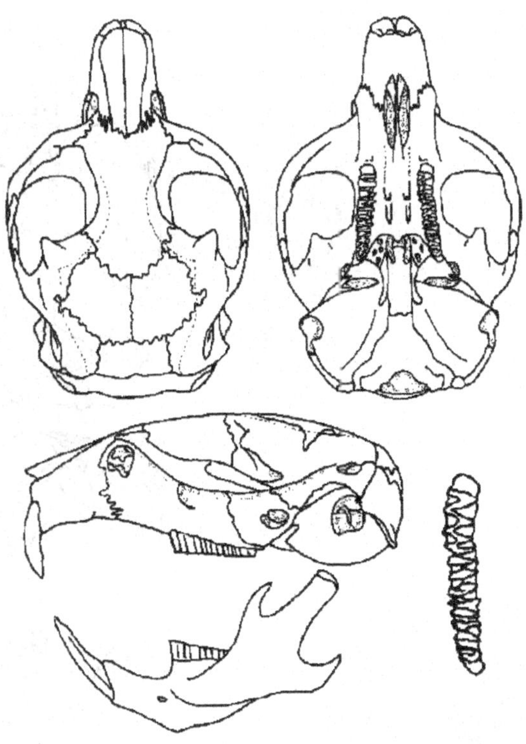

Fig. 62. *Dicrostonyx*
Greatest length of skull 27mm

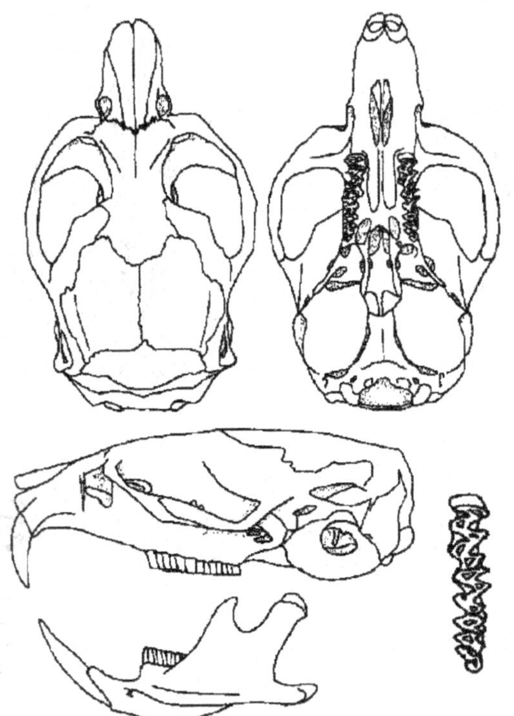

Fig. 63. *Microtus*
Greatest length of skull 30mm

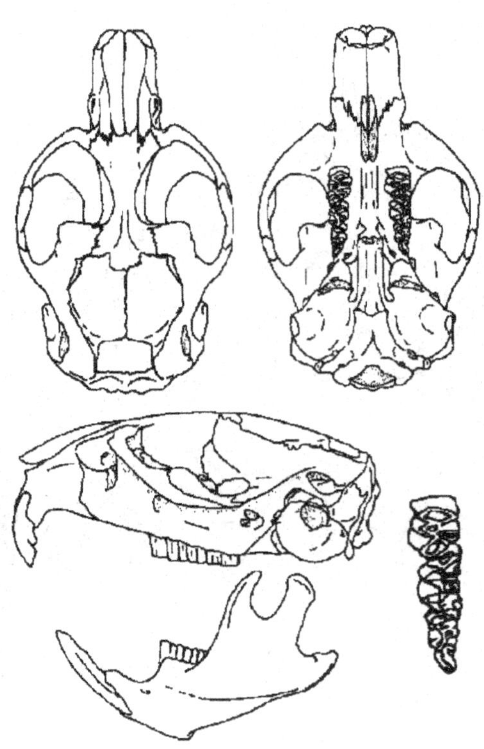

Fig. 64. *Neofiber*
Greatest length of skull 45mm

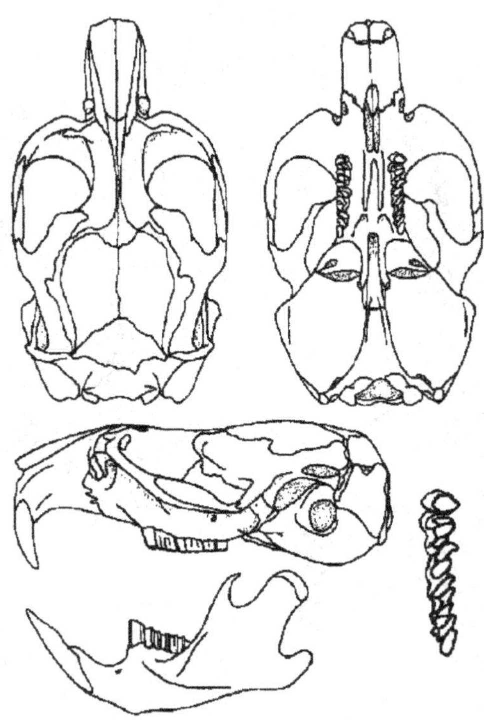

Fig. 65. *Lemmiscus*
Greatest length of skull 25mm

# LAGOMORPHA

1.    -Supraorbital processes absent; jugal projecting back almost to auditory meatus, forming a long spine; fenestrae on side of rostrum not covered with bony latticework; dental formula I2/1 C0/0 P3/2 M2/3 ... Ochotonidae.....*Ochotona* (Fig. 66)
-Supraorbital processes present, divided into antorbital and postorbital portions; jugal extending only approximately halfway to auditory meatus; fenestrae in side of rostrum covered by bony latticework; dental formula I2/1 C0/0 P3/2 M3/3 ...................................................... Leporidae......... 2

2.    -Interparietal absent in adults; supraorbital processes broadly triangular and winglike .....................................................*Lepus* (Fig. 67)
-Interparietal distinct in adults; supraorbital processes narrower and more strap-shaped ........................................................ 3

3.    -Supraorbital processes slender and rodlike, usually standing free from cranium; auditory bullae proportionately very large; anterior upper and lower premolar with only one fold in the enamel ................................. *Brachylagus* (Fig. 68)
-Supraorbital processes straplike, frequently touching or even fusing with cranium; auditory bullae proportionately much smaller; anterior upper and lower premolar with more than one fold; reentrant enamel crenate ............................................................................. 4

4.    -Interparietal rectangular in outline.............................................*Oryctolagus** (Fig. 69)
-Interparietal triangular in outline ...................................................*Sylvilagus* (Fig. 70)

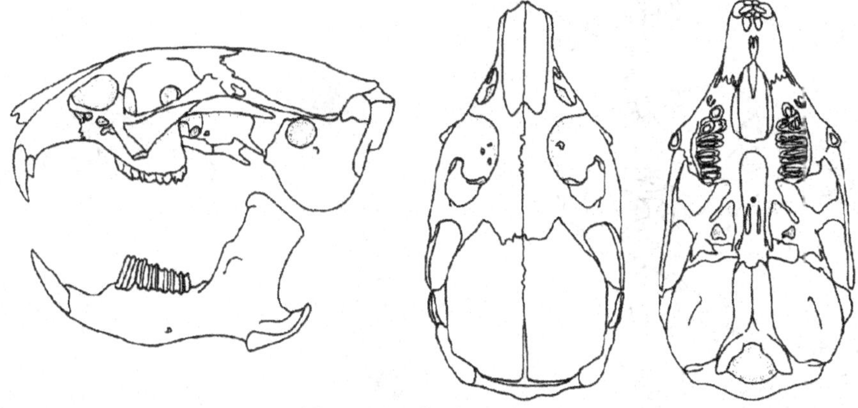

Fig. 66. *Ochotona*
Greatest length of skull 40mm

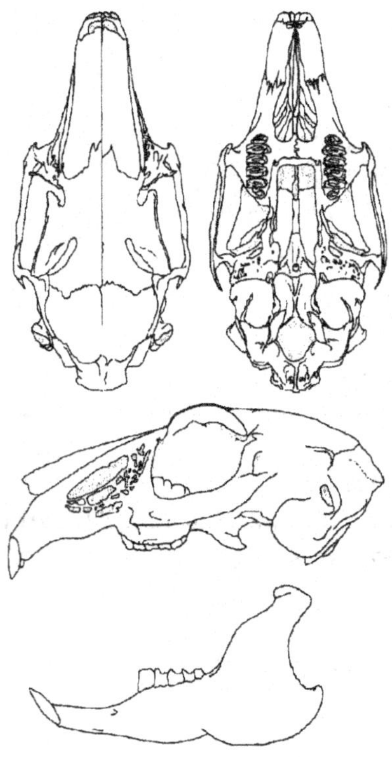

Fig. 67. *Lepus*
Greatest length of skull 95mm

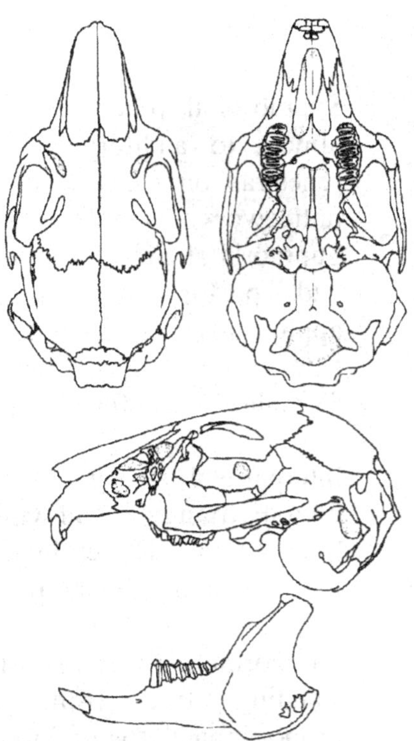

Fig. 68. *Brachylagus*
Greatest length of skull 50mm

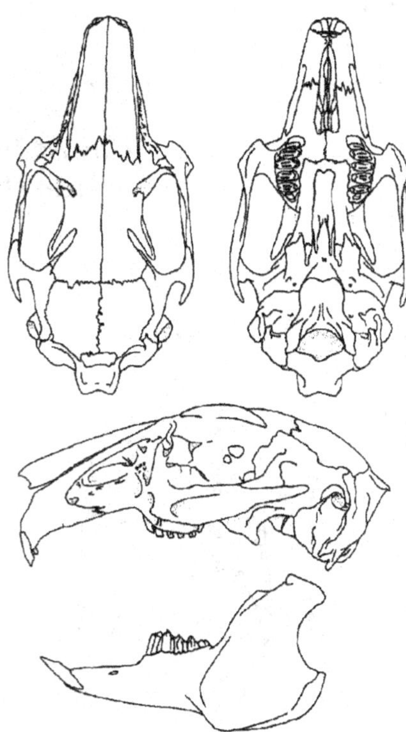

Fig. 69. *Oryctolagus**
Greatest length of skull 80mm

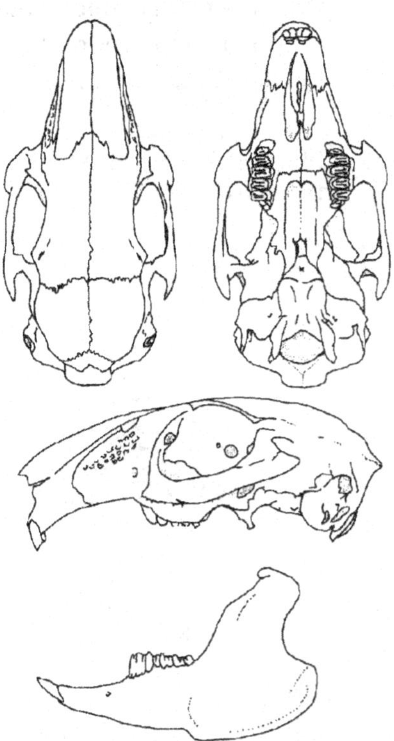

Fig. 70. *Sylvilagus*
Greatest length of skull 70mm

# SORICOMORPHA

1.  -Zygomatic arches complete; front pair of upper teeth moderately enlarged and projecting downward or only slightly forward, their longitudinal and lateral diameters approximately equal; teeth white ................................................................ Talpidae........ 2
    -Zygomatic arches incomplete; front pair of teeth greatly enlarged and projecting conspicuously forward, about five times as long as wide, and bearing a small posterior cusp; at least anterior teeth tipped reddish or blackish .......................... Soricidae ....... 6

2.  -Teeth totaling 36 (may be up to 40 in the young) ................................................. 3
    -Teeth totaling 44 (11 on each side above and below) ........................................... 4

3.  -Teeth 10 above and 8 below (10 below on each side in young); auditory bullae completely formed ...................................... Scalopus (Fig. 71)
    -Teeth 9 above and below on each side; auditory bullae incompletely formed ................................................................ Neurotrichus (Fig. 72)

4.  -Premaxillaries extending well forward below and in front of narial aperture; $I^1$ large and curved inward, $I^2$ minute, $I^3$ curved out and resembling a canine; rostrum long and slender; canines and first three premolars separated from adjacent teeth by spaces approximately equal to their own diameters; auditory bullae incompletely formed; posterior border of palate anterior to $M^3$; foramen magnum about twice as long as wide when seen from below ................................................ Condylura (Fig. 73)
    -Premaxillaries ending at narial aperture; incisor teeth more or less perpendicular; rostrum not conspicuously elongate; no teeth separated by noticeable spaces; auditory bullae complete or incomplete; posterior border of palate even with or behind $M^3$; foramen magnum 1-1.5 times as long as wide ................................................................................ 5

5.  -First upper incisor short, broad and flat in front with a distinct accessory cusp; auditory bullae incompletely formed ................................................................ Parascalops (Fig. 74)
    -First upper incisor long and broad, convex in front and flat behind; auditory bullae complete .......................................... Scapanus (Fig. 75)

6.  -Upper teeth 8 or 9 on each side; less than 5 unicuspids above; total of 28 or 30 teeth (Fig. 76) ........................................................... 7
    -Upper teeth 10 on each side; five unicuspids in each upper jaw behind enlarged anterior teeth; total of 32

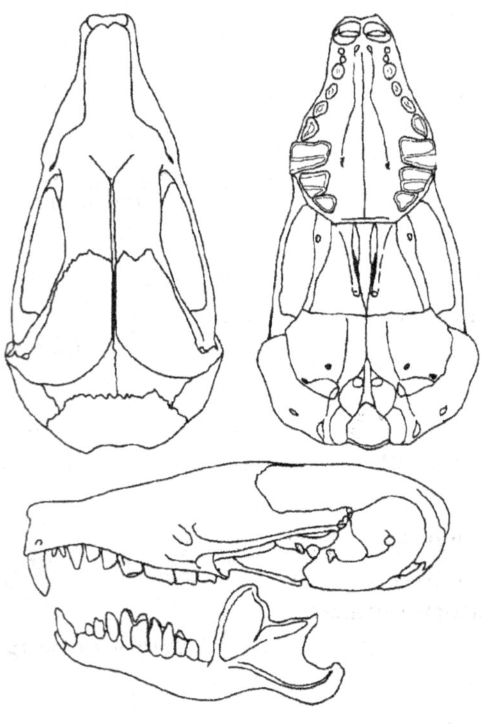

Fig. 71. *Scalopus*
Greatest length of skull 36mm

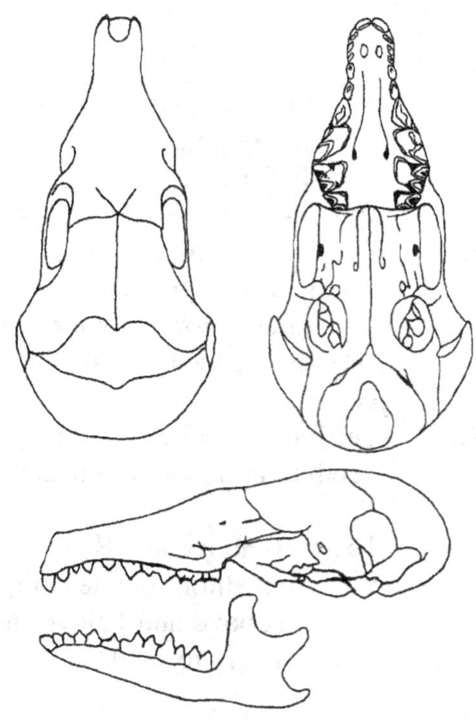

Fig. 72. *Neurotrichus*
Greatest length of skull 36mm

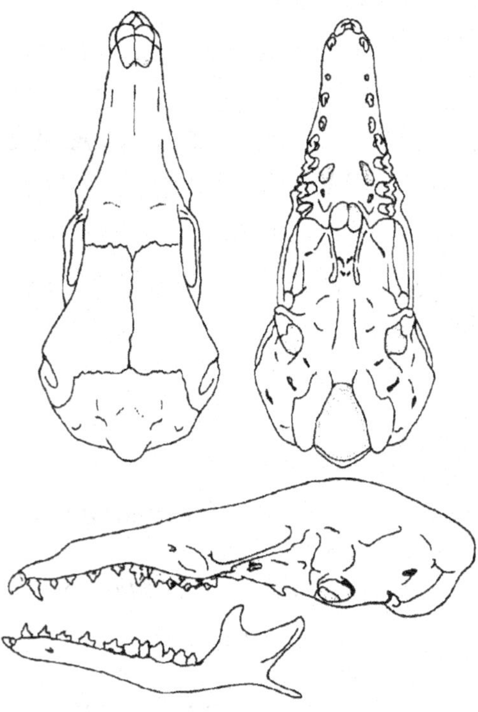

Fig. 73. *Condylura*
Greatest length of skull 35mm

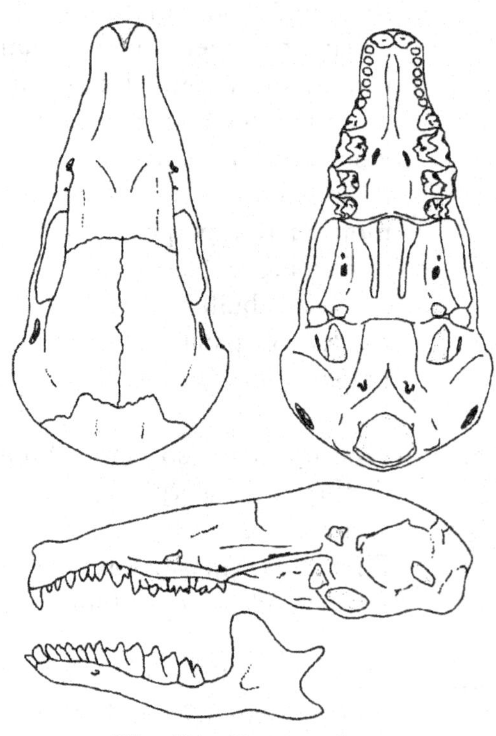

Fig. 74. *Parascalops*
Greatest length of skull 32mm

teeth (Fig. 76) ............................................................................................................. 8

7.       -Upper teeth 8 on each side, 3 unicuspids above, total of
28 teeth; teeth white except for enlarged anterior tooth
and anterointernal cusp of first large cheektooth ........................... *Notiosorex* (Fig. 77)
-Upper teeth 9 on each side, 4 unicuspids above, total of
30 teeth; all teeth except last unicuspid pigmented ........................... *Cryptotis* (Fig. 78)

8.       -Skull robust and strong, bones heavy; a distinct sagittal
crest present in mature specimens; first two unicuspids
of identical size, the third and fourth identical and much
smaller than the first two, the fifth minute and nearly or
quite obscured from lateral view by a projection of $P^4$ ...................... *Blarina* (Fig. 79)
-Skull delicate, bones thin and light; no sagittal crest;
first four upper unicuspids not arranged in two distinct
pairs of quite different sizes, sometimes almost
uniform; fifth unicuspid minute but clearly visible from
lateral view ............................................................................................................. 9

9.       -Third unicuspid similar in size to the second and fourth;
first and second unicuspids without an accessory inner
cusp ................................................................................................ *Sorex* (Fig. 80)
-Third upper unicuspid minute and wedged between the
second and fourth so as to be virtually invisible from
the side; first and second unicuspids with an accessory
inner cusp ................................................................... *Sorex* (*Microsorex*) (Fig. 81)

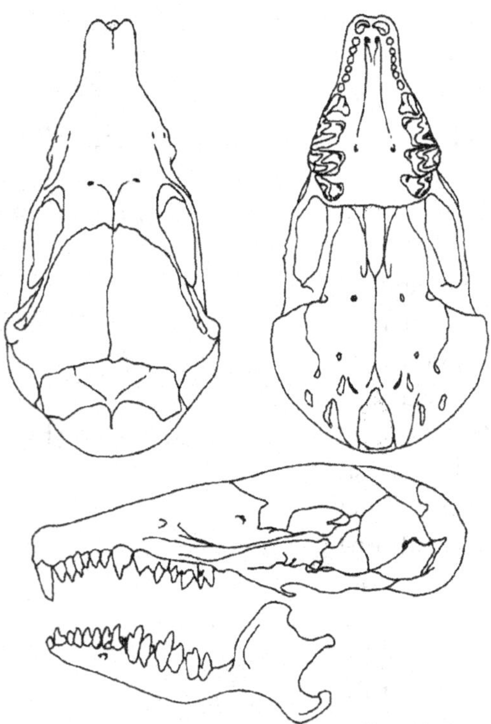
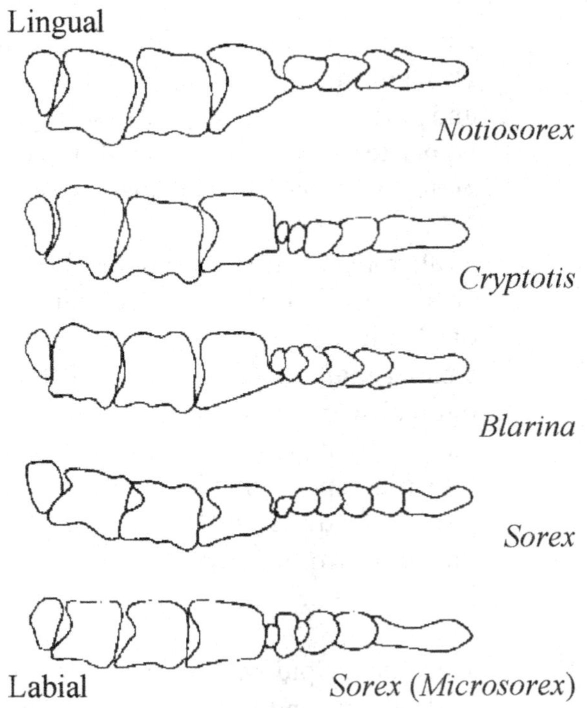


Fig. 75 *Scapanus*
Greatest length of skull 34mm

Fig. 76. Comparative occlusal arrangements of left upper tooth rows in Soricids (Not drawn to scale).

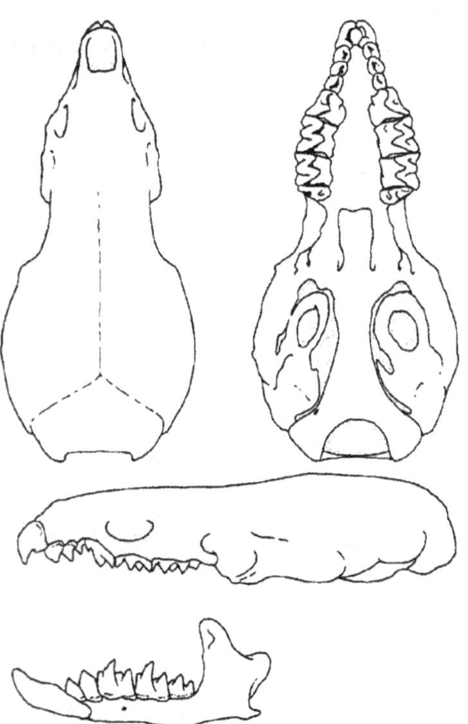

Fig. 77. *Notiosorex*
Greatest length of skull 17mm

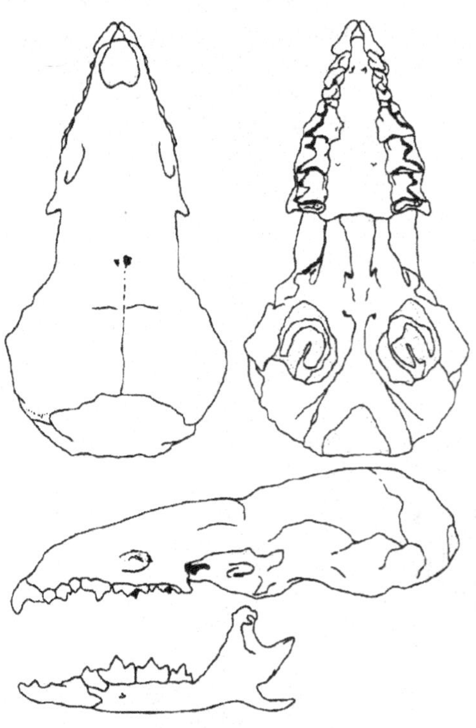

Fig. 78. *Cryptotis*
Greatest length of skull 16mm

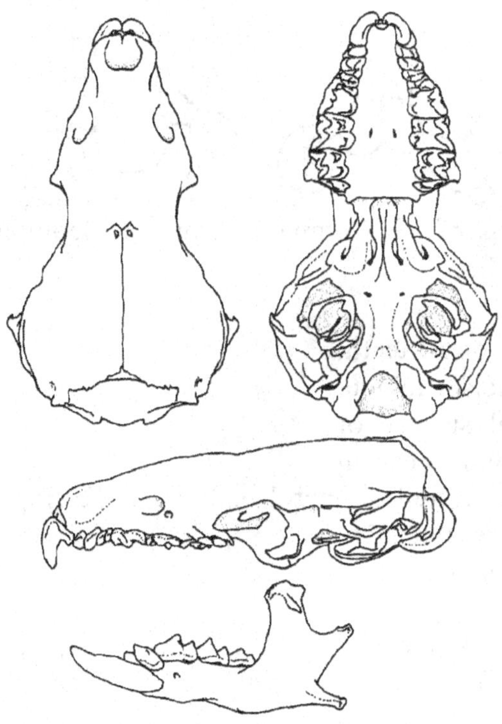

Fig. 79. *Blarina*
Greatest length of skull 24mm

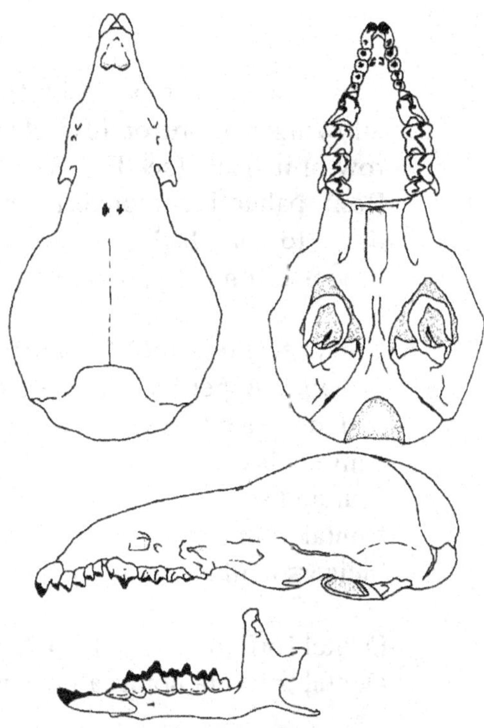

Fig. 80. *Sorex*
Greatest length of skull 16mm

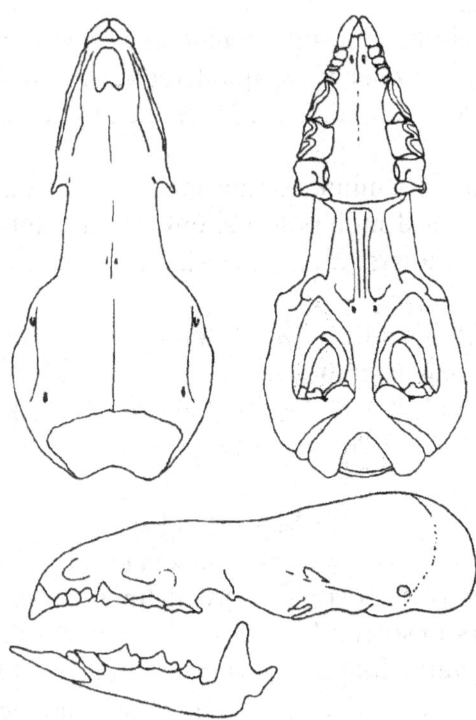

Fig. 81. *Sorex (Microsorex)*
Greatest length of skull 15mm

1.    -Bony palate complete in front allowing a more or less closed row of incisors (see Fig. 82) ........ 2
    -Bony palate incomplete in front due to a lack of palatal processes on the premaxillae ....... 9

Fig. 82. L Complete palate, R Incomplete palate

2    -Dental formula I2/2 C1/1 P1/2 M2/2, total of 26 teeth; principal upper incisor bladelike; second incisor minute and displaced inward, lying along medial surface of canine; lower incisors with serrated margins bearing four and seven cusplets ...................................Phyllostomatidae...... *Diphylla* (Fig. 83)
    -Dental formula not as above; teeth more than 26; incisors neither bladelike nor modified as above........................................................................... 3

3.    -Dental formula I2/2 C1/1 P2/3 M3/3, total of 34 teeth ..................................................... 4
    -Dental formula not as above, teeth less than 34......................Phyllostomatidae......... 5

4.    -Skull short, rostrum and braincase as broad or broader than long; forehead vertical; entire braincase so elevated that base of foramen magnum is above level of top of rostrum...................................................... Mormoopidae .... *Mormoops* (Fig. 84)
    -Skull slender and light, rostrum narrow and tapering; braincase sloping at forehead; occipital region marked off by a constriction .....................................Phyllostomatidae..... *Macrotus* (Fig. 85)

5.    -Incisors 2/2; third molars minute or absent................................................................. 6
    -Incisors 1/2 or 2/0; third molars large, only moderately smaller than second molars ................................................................................................ 7

6.    -Dental formula I2/2 C1/1 P2/3 M2/2, total of 30 teeth; rostrum at least as long as braincase ...........................................*Leptonycteris* (Fig. 86)
    -Dental formula I2/2 C1/1 P2/2 M2/3, total of 30 teeth; rostrum short, approximately half as long as braincase ....................*Artibeus* (Fig. 87)

7.    -Dental formula I2/0 C1/1 P2/3 M3/3, total of 30 teeth; upper incisors minute, closely crowded, separated from canines and incisors of opposite side by distinct spaces; W pattern of molars obsolete; braincase elevated above rostrum; rostrum very long and narrow; zygomata incomplete....................................Phyllostomatidae........*Choeronycteris* (Fig. 88)
    -Dental formula I1/1-2 C1/1 P2/2 M3/3, total of 28 or 30 teeth; upper incisors large and in contact; W pattern of

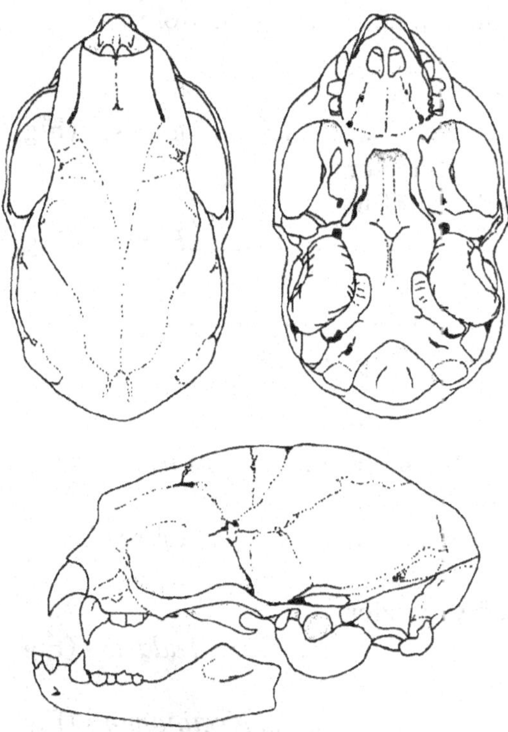

Fig. 83.  *Diphylla*
Greatest length of skull 23mm

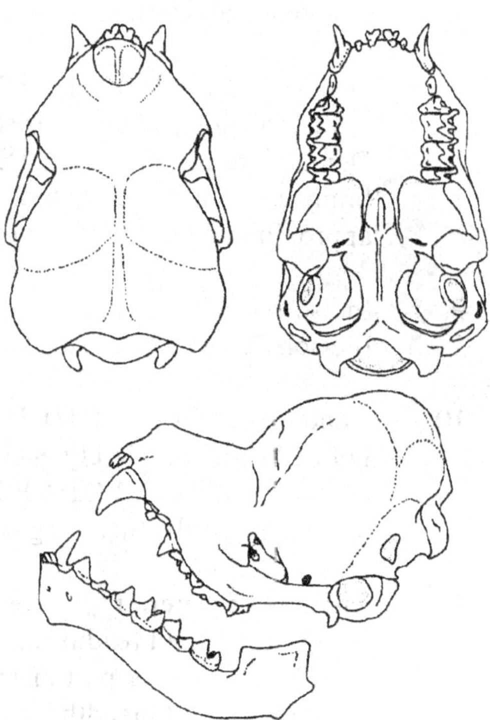

Fig. 84.  *Mormoops*
Greatest length of skull 14mm

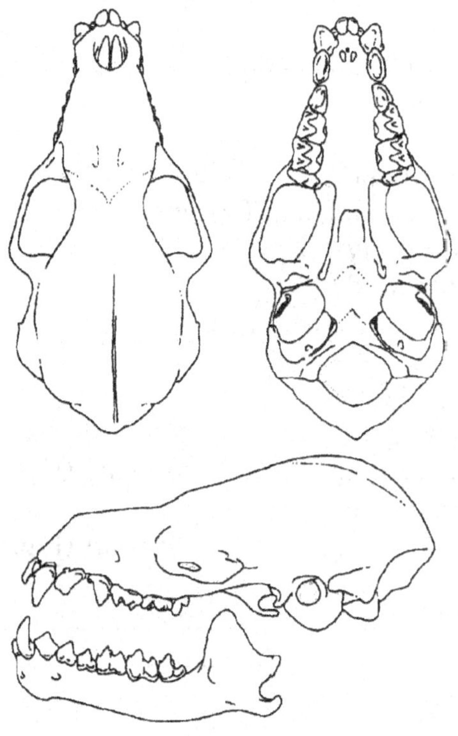

Fig. 85.  *Macrotus*
Greatest length of skull 22mm

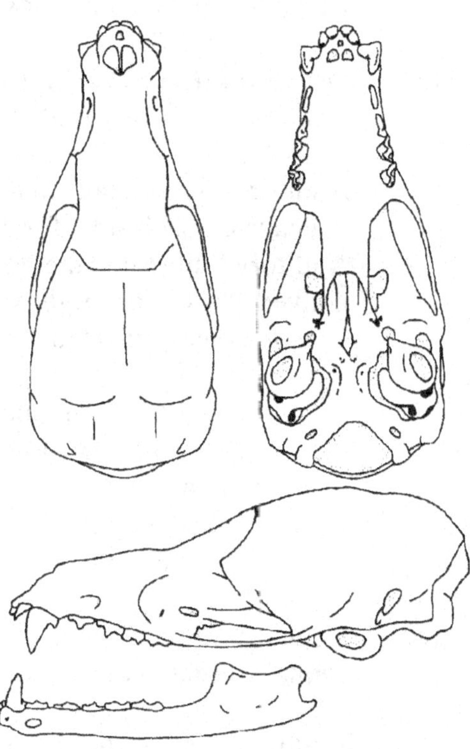

Fig. 86.  *Leptonycteris*
Greatest length of skull 27mm

molars well developed; rostrum short and wide;
zygomata complete ............................................................ Molossidae ......... 8

8.     -Dental formula I1/1 C1/1 P2/2 M3/3, total of 28 teeth;
braincase elevated above rostrum ...................................... *Molossus* (Fig. 89)
-Dental formula I1/2 C1/1 P2/2 M3/3, total of 30 teeth;
braincase not elevated; entire dorsal profile
approximately a straight line ............................................. *Eumops* (Fig. 90)

9.     -Incisors 1/2 ................................................................................ 10
-Incisors 2/3 or 1/3 ................................................................... 12

10.    -Dental formula I1/2 C1/1 P1/2 M3/3, total of 28 teeth;
bulla almost completely enclosing cochlea ....... Vespertilionidae .... *Antrozous* (Fig. 91)
-Dental formula I1/2 C1/1 P 2/2 M3/3, total of 30 teeth;
bulla only partly enclosing cochlea ..................................... Molossidae ....... 11

11.    -Breadth of anterior part of rostrum markedly greater
than interorbital breadth .................................................. *Tadarida* (Fig. 92)
-Breadth of anterior part of rostrum only slightly greater
than interorbital breadth .................................................. *Nyctinomops* (Fig. 93)

12.    -Dental formula I2/3 C1/1 P3/3 M3/3, total of 38 teeth
(occasionally $P^3$ may be crowded out of line or missing) ..................... *Myotis* (Fig. 94)
-Dental formula not as above, teeth less than 38 ...................................... 13

13.    -Dental formula I2/3 C 1/1 P2/3 M3/3, total of 36 teeth ............................... 14
-Dental formula not as above, teeth less than 36 ...................................... 16

14.    -Auditory bullae not much enlarged; rostrum broad,
concave on each side; forehead nearly flat ............................... *Lasionycteris* (Fig. 95)
-Auditory bullae much enlarged; rostrum narrow, evenly
convex above, or slightly concave medially; forehead
conspicuously elevated ................................................................... 15

15.    -Rostrum with pronounced lateral process above
infraorbital canal; breadth of braincase amounting to
more than half of greatest length of skull ................................... *Idionycteris* (Fig. 96)
-Rostrum lacking lateral process; breadth of braincase
amounting to less than half of greatest length of skull .............. *Corynorhynus* (Fig. 97)

16.    -Dental formula I2/3 C1/1 P2/2 M3/3, total of 34 teeth ............................... 17
-Dental formula not as above, less than 34 teeth ...................................... 19

17.    -Auditory bullae large and elongate, emarginate on
median border; canines small and weak, lower ones

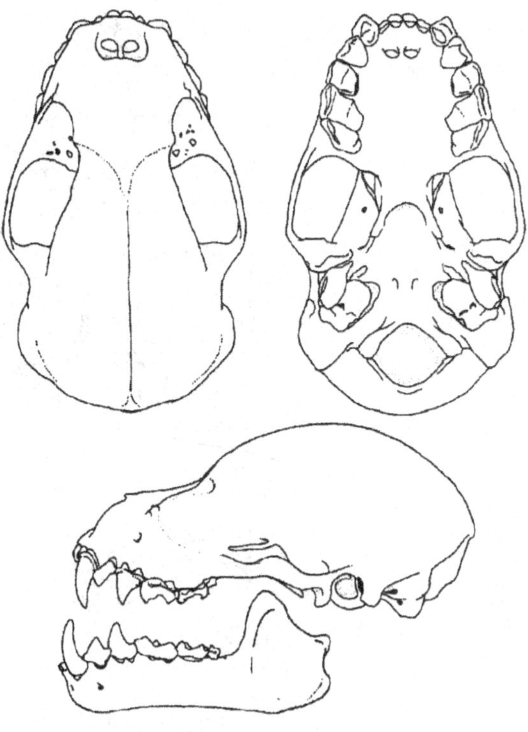

Fig. 87. *Artibeus*
Greatest length of skull 28mm

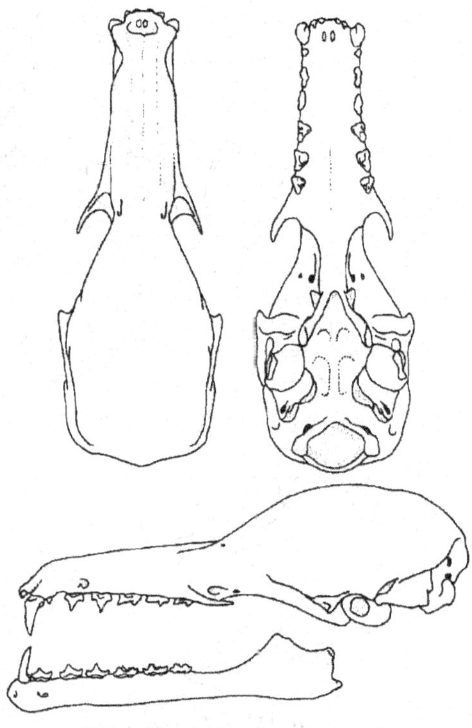

Fig. 88. *Choeronycteris*
Greatest length of skull 30mm

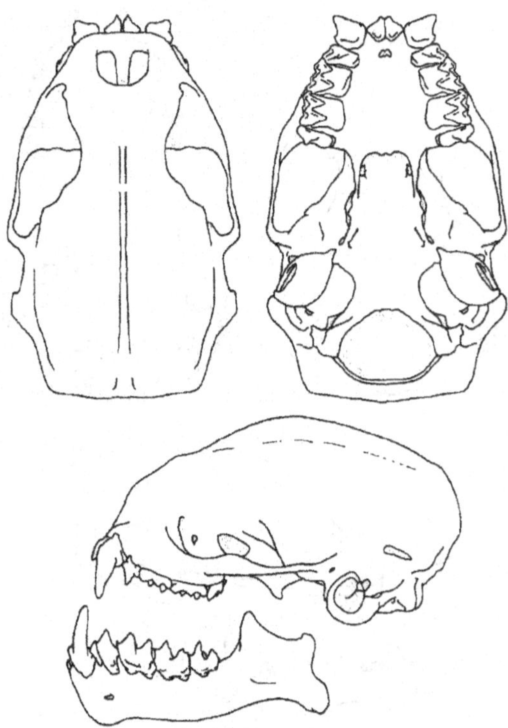

Fig. 89. *Molossus*
Greatest length of skull 17mm

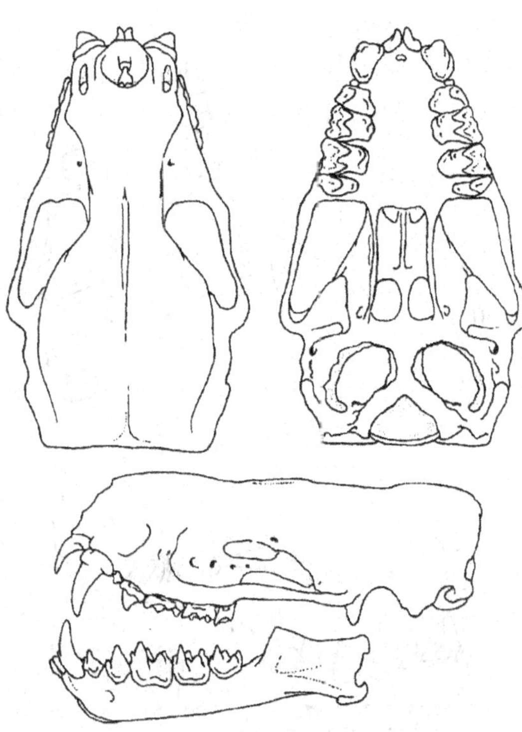

Fig. 90. *Eumops*
Greatest length of skull 28mm

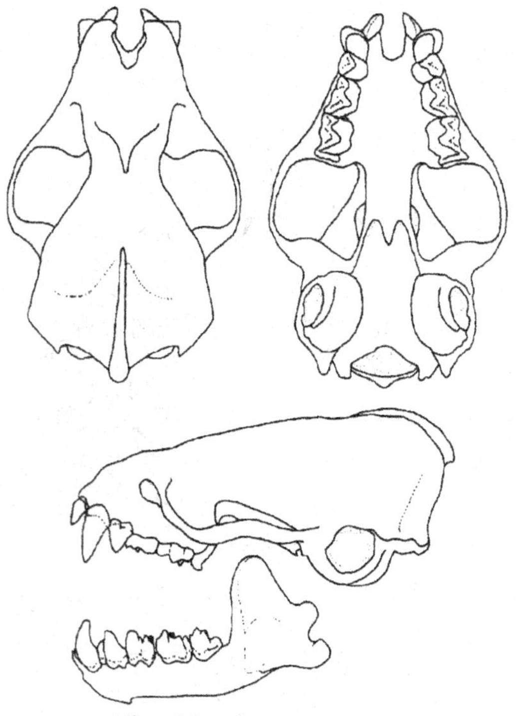

Fig. 91. *Antrozous*
Greatest length of skull 21mm

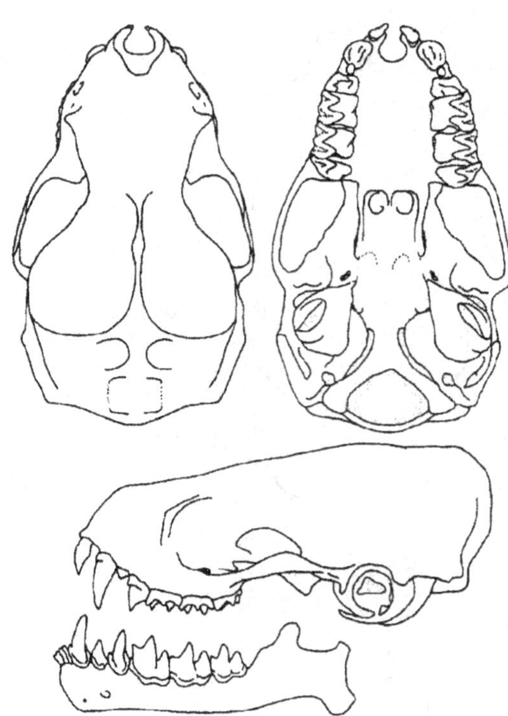

Fig. 92. *Tadarida*
Greatest length of skull 17mm

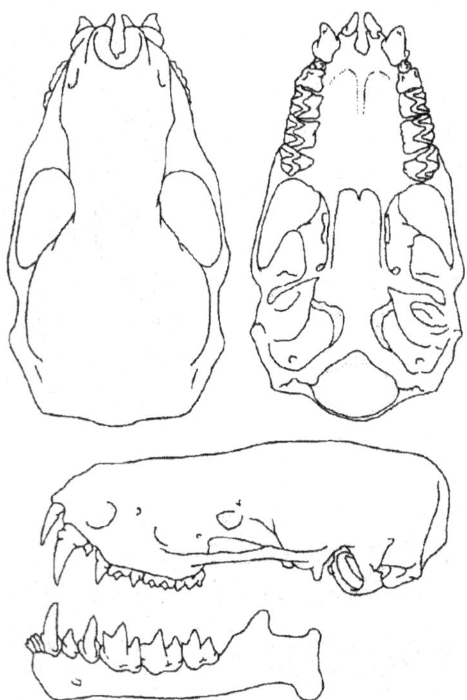

Fig 93. *Nyctinomops*
Greatest length of skull 20mm

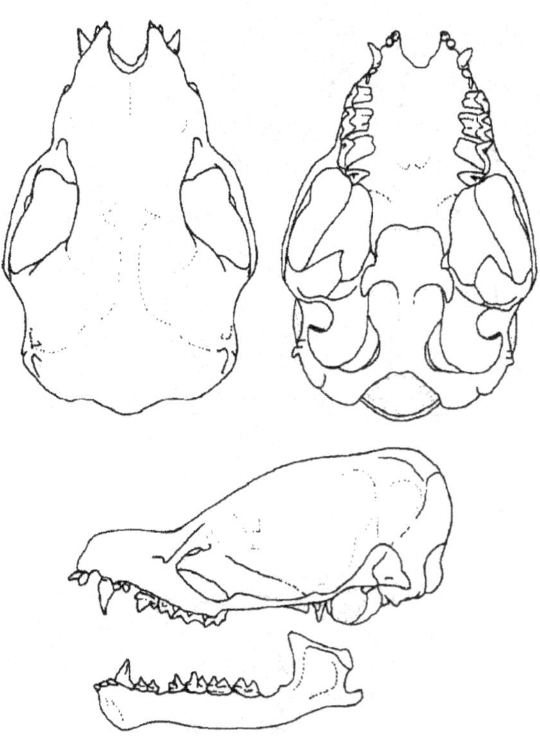

Fig. 94. *Myotis*
Greatest length of skull 15mm

with a distinct accessory cusp behind; zygomata abruptly widened in middle .................................................*Euderma* (Fig. 98)

-Auditory bullae not enlarged, elongate and emarginate; canines normal, unicuspidate; zygomata of uniform height throughout ................................................................................ 18

18.  -Skull nearly straight in dorsal profile, palate extends far behind molars ...................................................................*Parastrellus* (Fig. 99)

-Skull concave in dorsal profile, palate extends only slightly behind molars .................................................... *Perimyotis* (Fig. 100)

19.  -Dental formula I1/3 C1/1 P1/2 M3/3, total of 30 teeth ............................................... 20

-Dental formula not as above, total of 32 teeth ............................................................ 21

20.  -Upper incisor in contact with canine; third lower incisor crowded and smaller than first and second incisors..... *Lasiurus* (*Dasypterus*; Fig. 101)

-Upper incisor separated from canine; third lower incisor not crowded and equal in size to first and second ........................*Nycticeius* (Fig. 102)

21.  -Dental formula I2/3 C1/1 P1/2 M3/3, dorsal profile straight and sloping upward from nares to occipital crest ................................................................................... *Eptesicus* (Fig. 103)

-Dental formula I1/3 C1/1 P2/2 M3/3 ................................................................... 22

22.  -Skull high and short, greatest height equal to half total length; greatest width of rostrum equaling greatest width of braincase .................................................... *Lasiurus* (*Lasiurus*; Fig. 104)

-Skull lower and more elongate, greatest height equal to less than half total length; greatest width of rostrum narrower than greatest width of braincase ......................................*Tadarida* (Fig. 92)

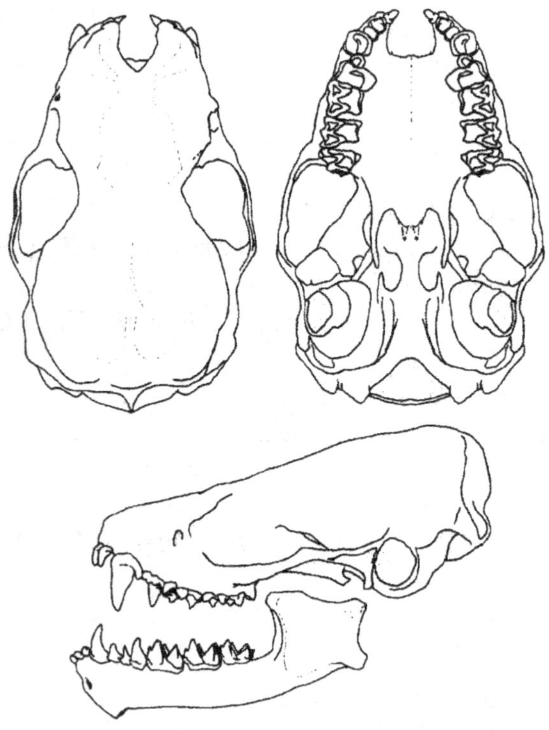

Fig. 95. *Lasionycteris*
Greatest length of skull 15mm

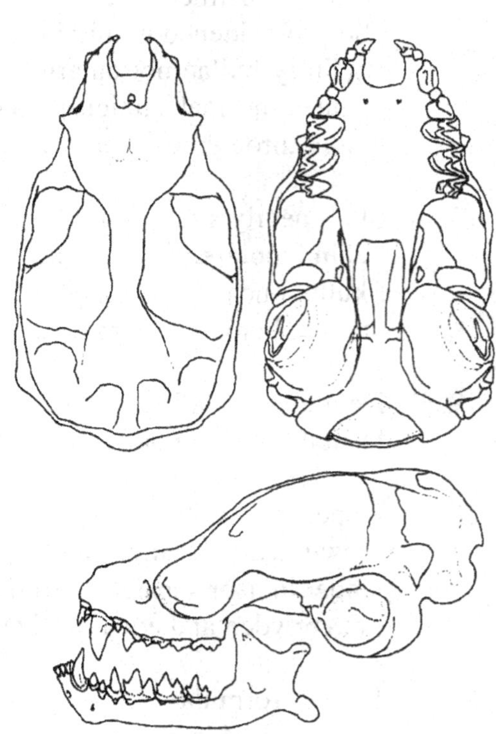

Fig. 96. *Idionycteris*
Greatest length of skull 17mm

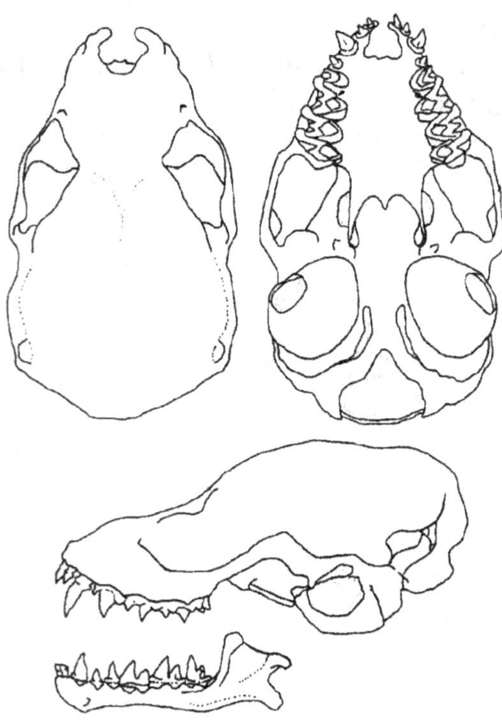

Fig. 97. *Corynorhynus*
Greatest length of skull 16mm

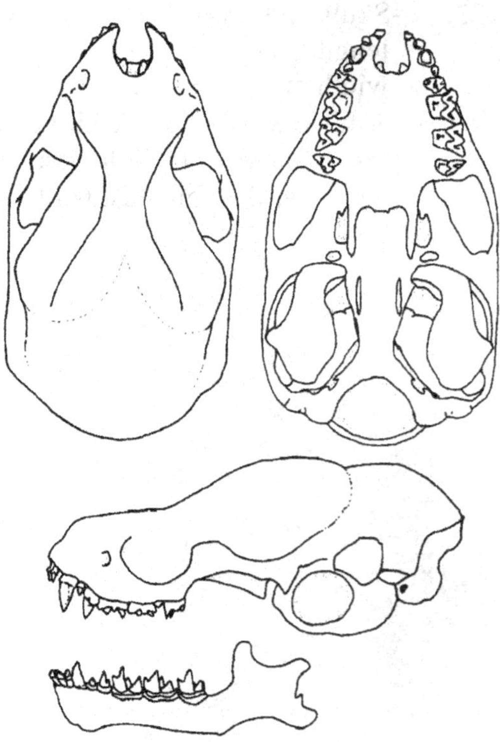

Fig. 98. *Euderma*
Greatest length of skull 19mm

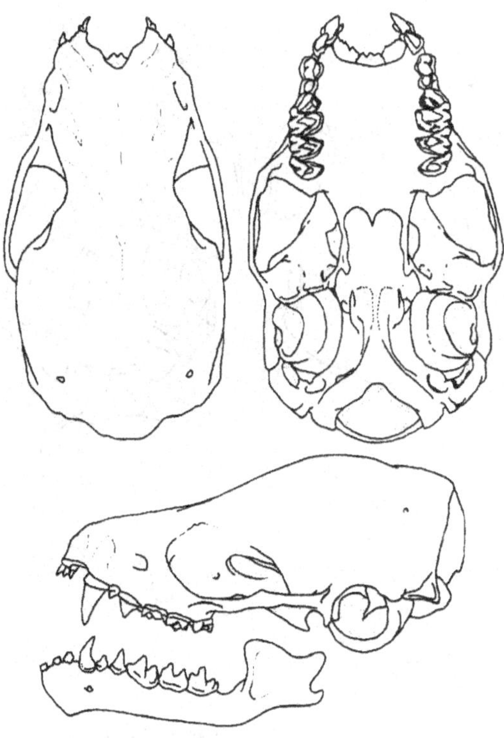

Fig. 99. *Parastrellus*
Greatest length of skull 15mm

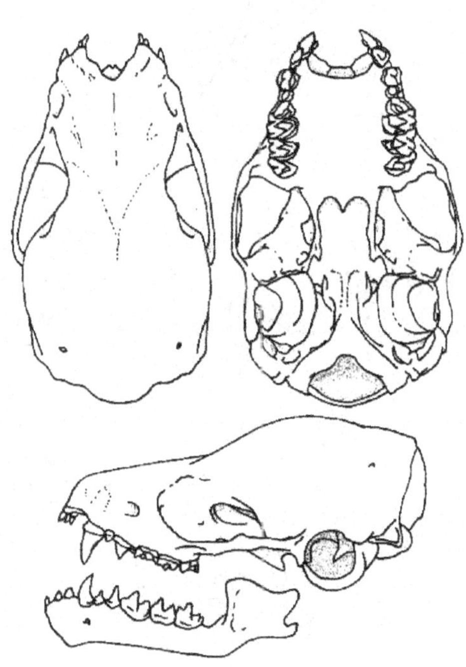

Fig. 100. *Perimyotis*
Greatest length cf skull 15mm

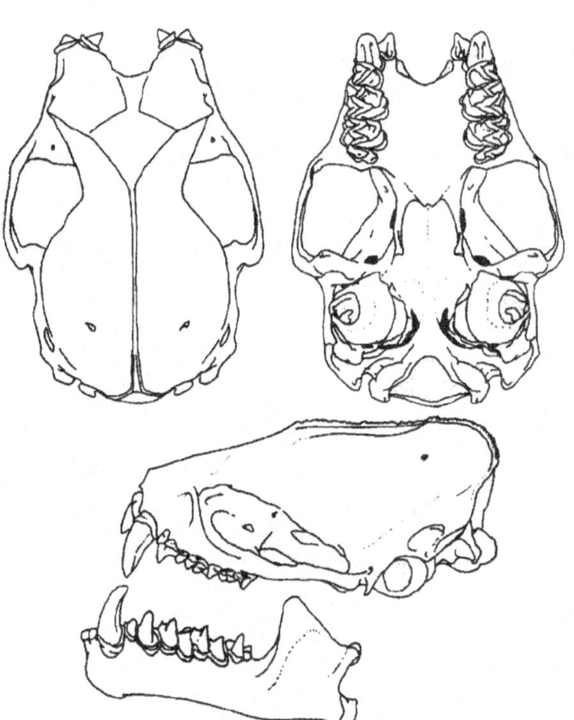

Fig. 101. *Lasiurus (Dasypterus)*
Greatest length of skull 17mm

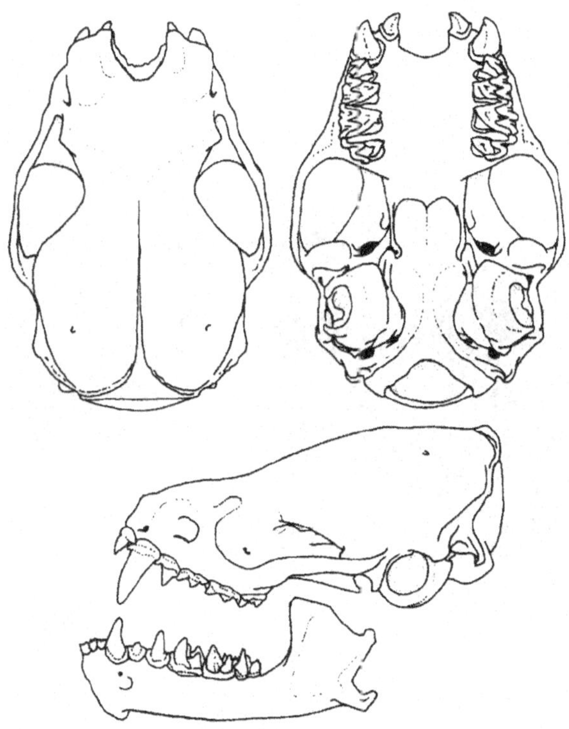

Fig. 102. *Nycticeius*
Greatest length of skull 14mm

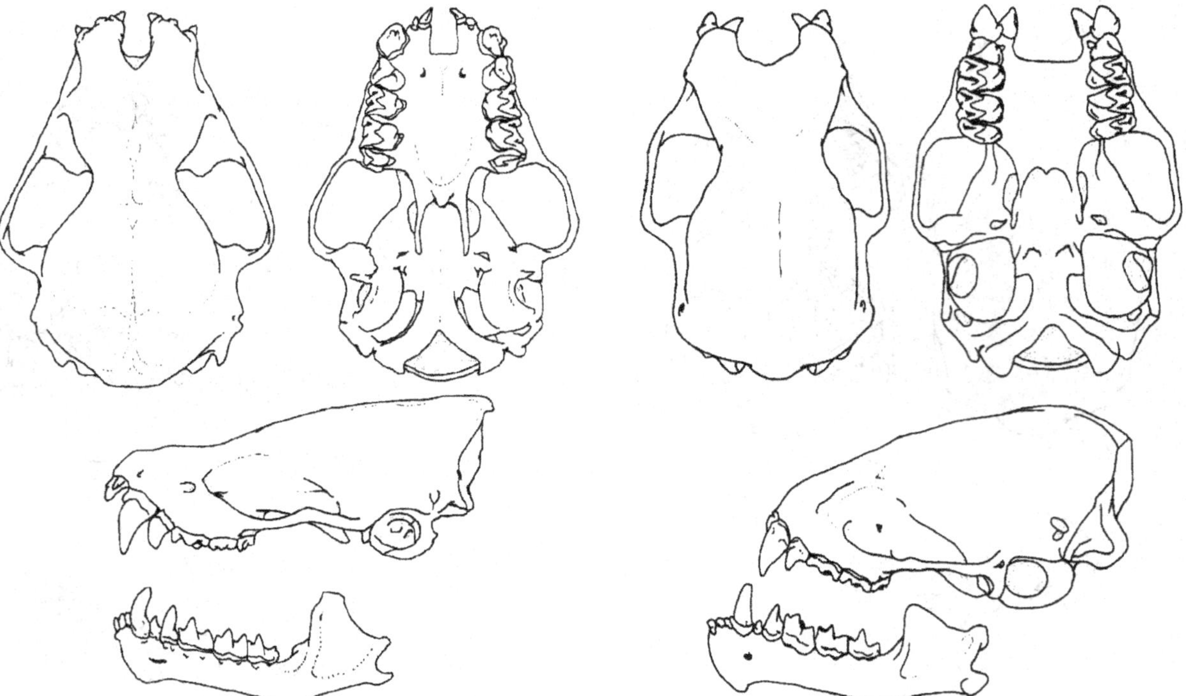

Fig. 103. *Eptesicus*
Greatest length of skull 20mm

Fig. 104. *Lasiurus* (*Lasiurus*)
Greatest length of skull 13mm

# CARNIVORA

1.    -Jaw teeth not all alike, but adapted for shearing (carnassial) or grinding (bunodont) ......................................Suborder Fissipedia......... 2
    -Jaw teeth simple and all very much alike, being either simple cones, cones with accessory cusps, or flat and buttonlike; upper incisors 3, 2, or 1 in number ...................Suborder Pinnipedia....... 28

2.    -Rostrum much shortened, entire top of head evenly convex; jaw teeth shearing, without grinding surfaces; molars 1/1, upper molar minute; total of 28 or 30 teeth ..........................Felidae......... 3
    -Rostrum not greatly shortened, or if so, entire top of head not evenly convex; teeth shearing or grinding; molars never 1/1; total of 32 or more teeth (bears often have some or all of first 3 upper and lower premolars lost) ....................................................................................................... 8

3.    -Total length of skull greater than 158mm................................................. 4
    -Total length of skull less than 158mm ..................................................... 5

4.    -Bregmatic processes of parietals extending antero-medially over frontals, approaching or reaching temporal crest.....................................................................*Puma* (*Puma*; Fig. 105)
    -Bregmatic processes of parietals not as above.................................*Panthera* (Fig. 106)

5.    -Premolars 2/2; total of 28 teeth ..............................................*Lynx* (Fig. 107)
    -Premolars 3/2; total of 30 teeth ............................................................. 6

6.    -Top of skull only slightly arched; nasals extending beyond anterior margin of palatine foramen; palatine foramen not visible in the dorsal view ............................ *Puma* (*Herpailurus*; Fig. 108)
    -Top of skull highly arched; nasals not extending beyond anterior margin of palatine foramen; palatine foramen visible in the dorsal view ................................................................ 7

7    -Length of $P^4$ more than 12.7mm ..................................*Leopardus* (Fig. 109)
    -Length of $P^4$ less than 12.7mm.......................................... *Felis** (Fig. 110)

8.    -Dental formula I3/3 C1/1 P4/4 M2/3, total of 42 teeth (except sometimes in bears as noted above)............................................... 9
    -Dental formula not as above, total of less than 42 teeth ...................... 12

9.    -Size very large, usually exceeding 300mm in length; rostrum broad and massive; molar teeth broad and flat, no specialized carnassials ........................... Ursidae... *Ursus* (Fig. 111)....... 10

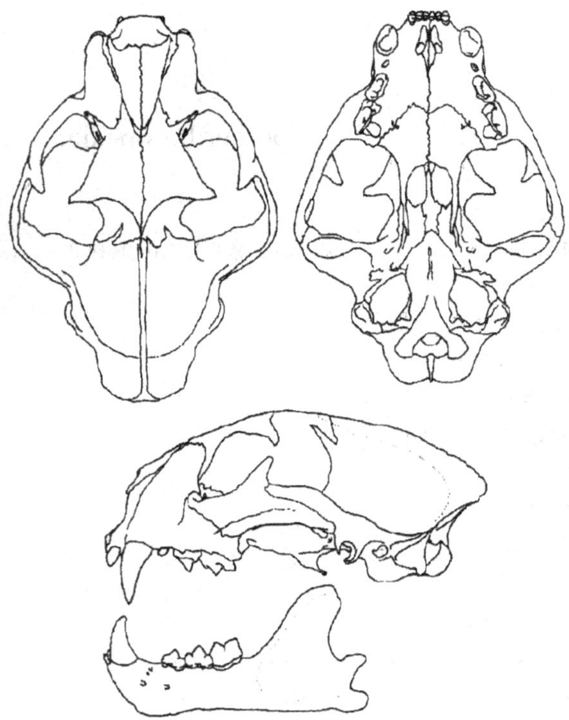

Fig. 105. *Puma* (*Puma*)
Greatest length of skull 200mm

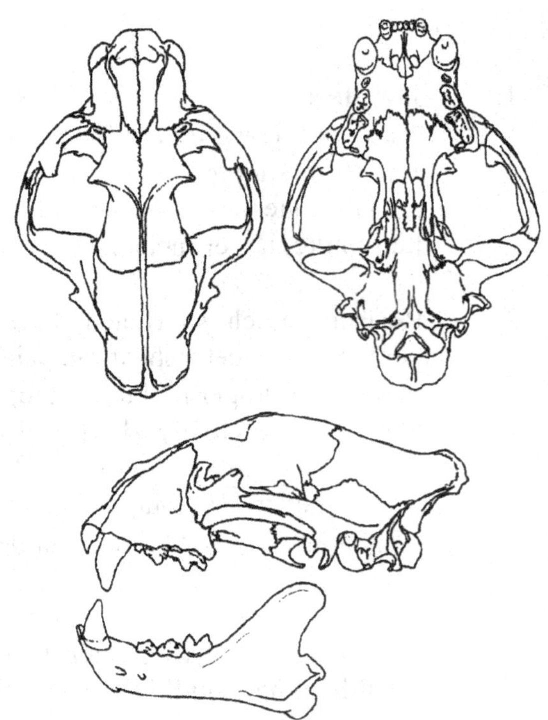

Fig. 106. *Panthera*
Greatest length of skull 250mm

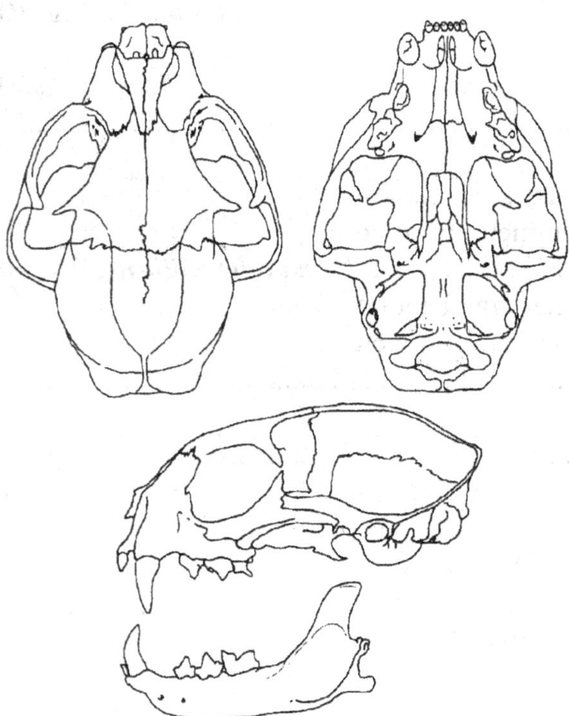

Fig. 107. *Lynx*
Greatest length of skull 130mm

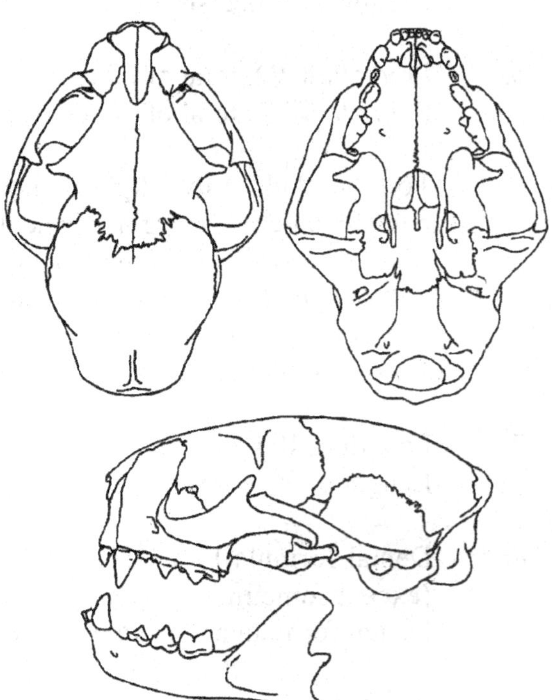

Fig. 108. *Puma* (*Herpailurus*)
Greatest length of skull 110mm

-Size moderate, rarely attaining 300mm; rostrum distinctly narrowed; carnassials fully developed .................................. Canidae....... 25

10. -Combined length of $M^1$ and $M^2$ less than breadth of palate; frontal region less elevated, giving a relatively flat dorsal profile; occipital condyles visible from above; $P^4$ more carnassial in appearance; $M^1$ and $M^2$ with more trenchant cusps and less sculptured crushing surfaces; palate concave................................................. (*Thalarctos*)
-Combined length of $M^1$ and $M^2$ equal to breadth of palate; frontal region more arched; occipital condyles not visible from above; $P^4$ not at all carnassial in appearance; $M^1$ and $M^2$ sculptured and lacking cusps; palate flat................................................................................. 11

11. -$M^2$ broadest at midpoint and not more than 1.5 times length of $M^1$ ....................................................................... (*Euarctos*)
-$M^2$ broadest at anterior end and more than 1.5 times length of $M^1$ ......................................................................... (*Ursus*)

12. -Dental formula I3/3 C1/1 P4/4 M2/2, total of 40 teeth .................... Procyonidae....... 13
-Dental formula not as above, total of less than 40 teeth ...................Mustelidae....... 15

13. -$P^4$ and $M_1$ shearing carnassials; posterior border of palate opposite last molar.............................................. *Bassariscus* (Fig. 112)
-No carnassials, jaw teeth bunodont; posterior border of palate behind last molars................................................................. 14

14. -Canines bladelike, with sharp anterior and posterior edges; nasals upturned anteriorly; internal nares divided by a median bony septum ............................................... *Nasua* (Fig. 113)
-Canines oval in cross-section; nasals not upturned anteriorly; internal nares undivided ..............................*Procyon* (Fig. 114)

15. -Dental formula I3/3 C1/1 P4/4 M1/2, total of 38 teeth; palate extending beyond last molars................................................. 16
-Dental formula not as above, less than 38 teeth; palate various................................................................................... 17

16. -Skull heavily built with a heavy sagittal crest projecting posterior to occiput ........................................................ *Gulo* (Fig. 115)
-Skull more lightly built, sagittal crest absent or only slightly developed ............................................................ *Martes* (Fig. 116)

17. -Dental formula I3/3 C1/1 P4/3 M1/2, total of 36 teeth; rostrum very short, as wide or wider than long;

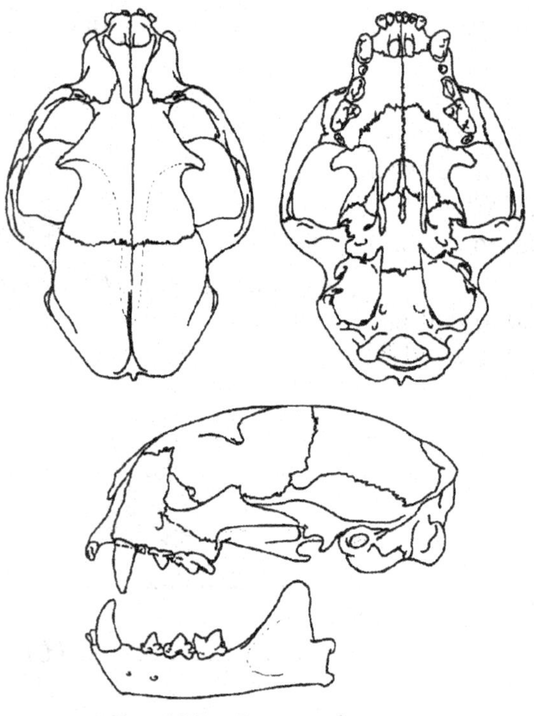

Fig. 109. *Leopardus*
Greatest length of skull 125mm

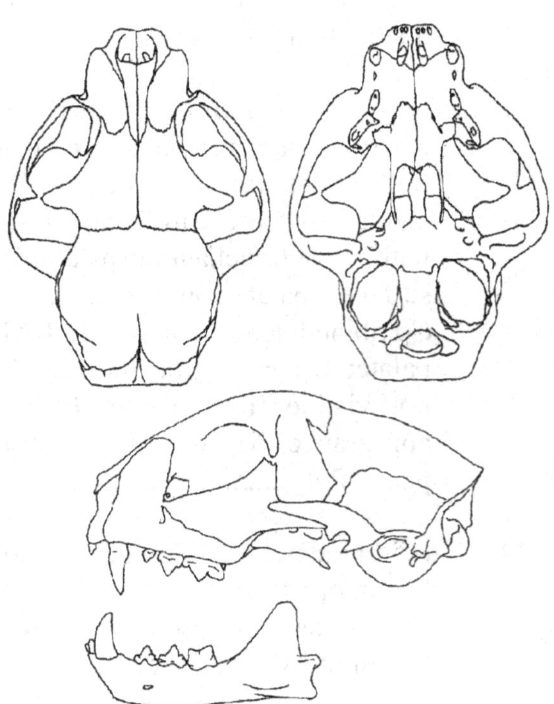

Fig. 110. *Felis**
Greatest length of skull 100mm

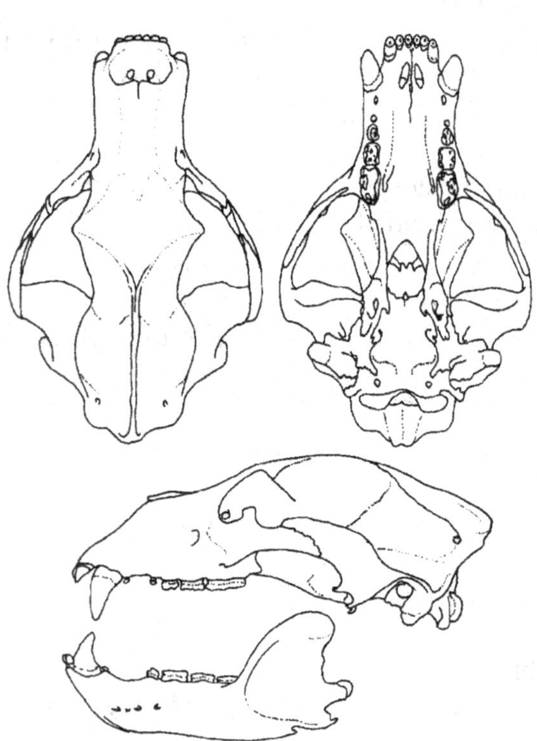

Fig. 111. *Ursus*
Greatest length of skull 360mm

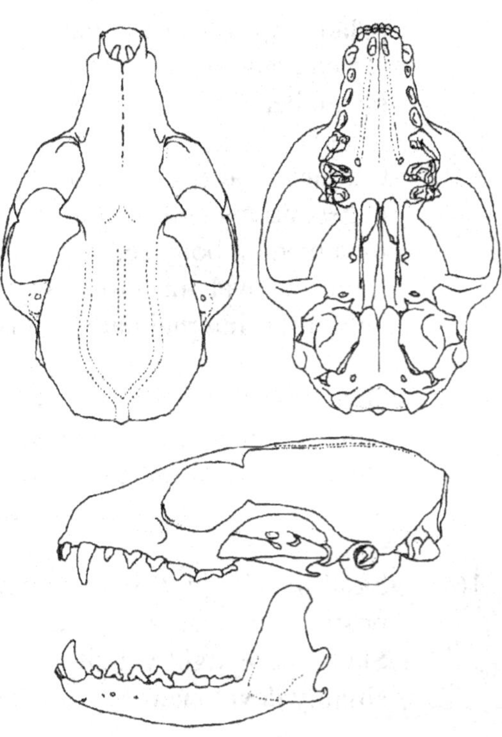

Fig. 112. *Bassariscus*
Greatest length of skull 70mm

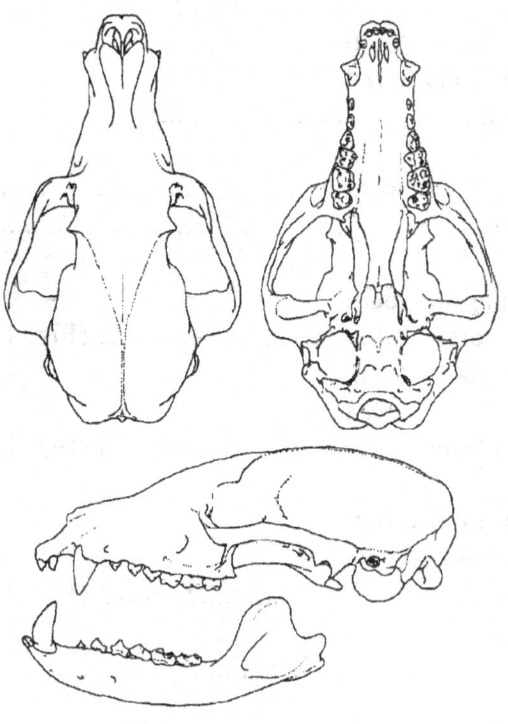

Fig. 113. *Nasua*
Greatest length of skull 125mm

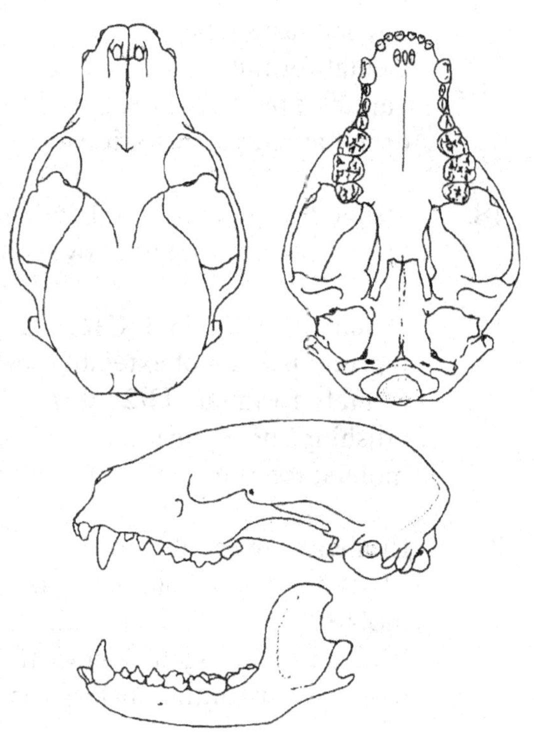

Fig. 114. *Procyon*
Greatest length of skull 120mm

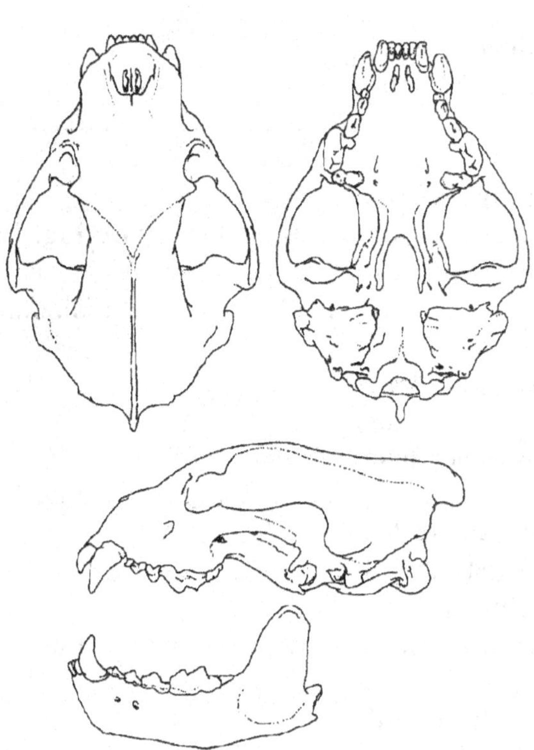

Fig. 115. *Gulo*
Greatest length of skull 140mm

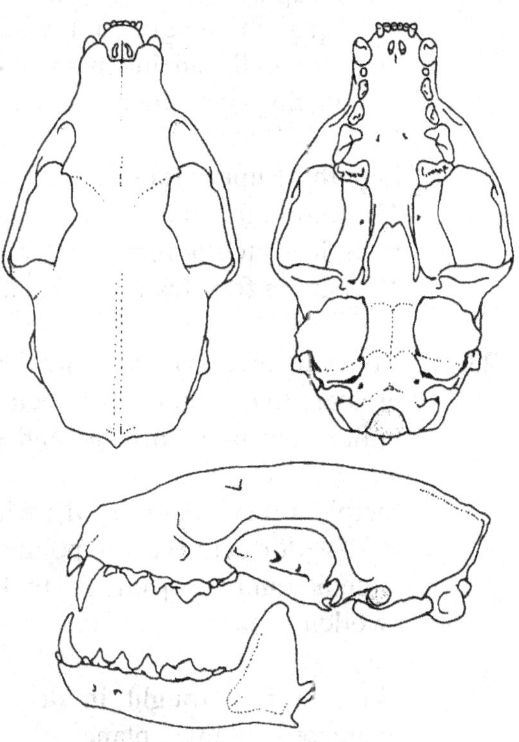

Fig. 116. *Martes*
Greatest length of skull 80mm

postorbital region much constricted; palate extending beyond last molars .................................................................*Lontra* (Fig. 117)
-Dental formula not as above, teeth less than 36; rostrum variable; postorbital region only moderately constricted or not at all; palate various ........................................................ 18

18. -Dental formula I3/3 C1/1 P3/3 M1/2, total of 34 teeth ................................................ 20
-Dental formula not as above, total of 32 teeth ........................................................ 19

19. -Dental formula I3/3 C1/1 P2/3 Ml/2, carnassial teeth present; palate not extending beyond last molars ......................... *Conepatus* (Fig. 118)
-Dental formula I3/2 C1/1 P3/3 M2/2, jaw teeth crushing, no carnassials; palate extending beyond last molars; rostrum very short, much wider than long ...........................*Enhydra* (Fig. 119)

20. -Braincase triangular in dorsal view, widest at occiput; $M^1$ triangular in outline; palate extending beyond last molars......................................................................................*Taxidea* (Fig. 120)
-Braincase elongate, not flaring behind; $M^1$ not triangular in outline; palate various ....................................................... 21

21. -$M^1$ wide, nearly twice as wide as long, outline somewhat dumbbell-shaped; auditory bullae nearly twice as long as wide and smoothly rounded; palate extending beyond last molars........................................................ 22
-$M^1$ squarish, length and width approximately equal; auditory bulla an inconspicuous rounded knob; palate terminating opposite last molars ........................................................ 24

22. -Length of upper tooth row less than 20mm in males or 17.8mm in females................................... *Mustela* (in part; Fig. 121)
-Length of tooth row greater than 20mm in males or 17.8mm in females................................................................... 23

23. -No trace of metaconid on $M_1$; width between canines greater than width between midpoints of auditory bullae; mastoids angular and projecting; bullae flatter and more circular in outline ................................................*Mustela* (Fig. 121)
-Incipient metaconid on $M_1$; width between canines less than width between midpoints of bullae; mastoids less angular and projecting; bullae more elongate and swollen ...............................................................*Neovison* (Fig. 122)

24. -Skull nearly straight in dorsal profile, rostrum little depressed below plane of frontal; mastoid region inflated, giving a flared appearance when viewed from

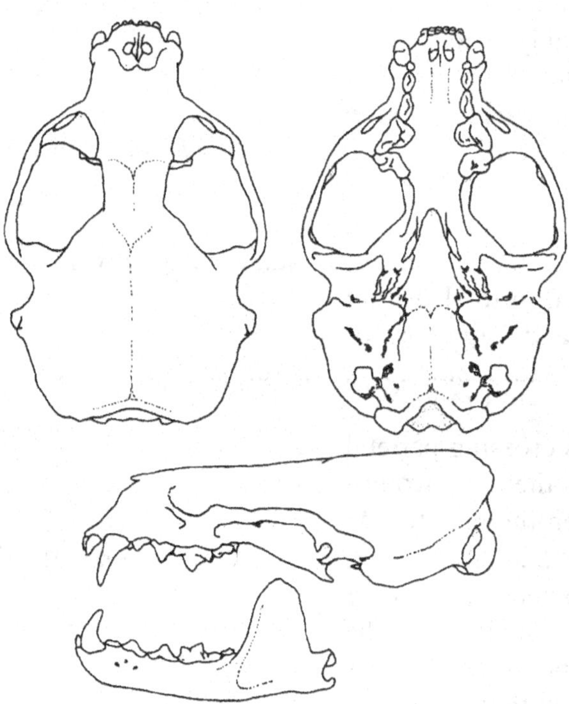

Fig. 117. *Lontra*
Greatest length of skull 125mm

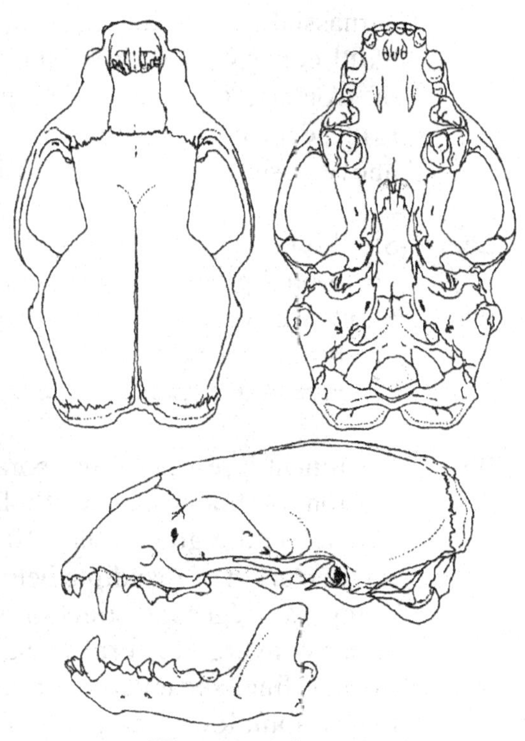

Fig. 118. *Conepatus*
Greatest length of skull 80mm

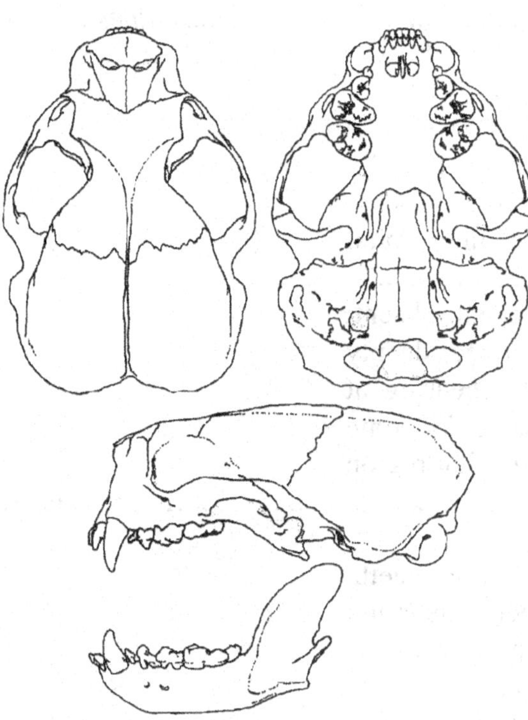

Fig. 119. *Enhydra*
Greatest length of skull 145mm

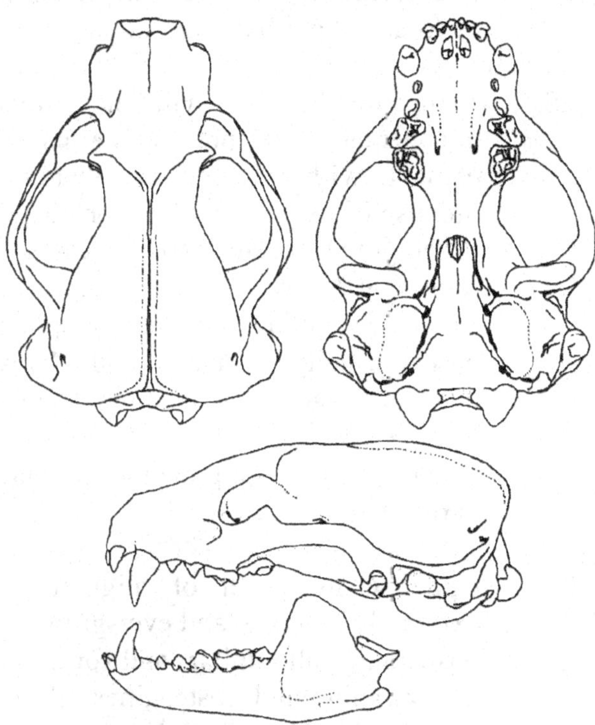

Fig. 120. *Taxidea*
Greatest length of skull 120mm

above; infraorbital canal opening above anterior half of carnassial ......................................................................... *Spilogale* (Fig. 123)
-Skull convex above, rostrum considerably below plane of frontals; mastoid region not inflated, that part of skull concave in outline; infraorbital canal opening above posterior half of carnassial ..................................... *Mephitis* (Fig. 124)

25.  -Skull relatively large, exceeding 150mm in total length; postorbital processes thick and convex ..................................... *Canis* (Fig. 125)
-Skull relatively small, less than 150mm in total length; postorbital processes of frontal thin, and depressed or concave above ............................................................................................... 26

26.  -Prominent lyre-shaped temporal ridges crossing parietal portion of braincase, width between them at fronto-parietal suture greater than 10mm; a prominent step on lower margin of mandible below ramus .............................. *Urocyon* (Fig. 126)
-Temporal ridges indistinct or absent, when present they either converge to form a sagittal crest or lie near the median line, separated at fronto-parietal suture by less than 10mm; lower margin of mandible without step ................................ 27

27.  -Interorbital region elevated; profile of forehead bulging in front of orbits ...................................................................... *Alopex* (Fig. 127)
-Interorbital region not conspicuously elevated; profile of forehead not bulging ................................................................ *Vulpes* (Fig. 128)

28.  -Tympanic bullae small, not smooth and swollen; alisphenoid canal present; nasals short and in normal contact with frontals; postcanine teeth essentially unicuspid, exclusive of minor cusplets that may occur on cingulum; postorbital processes present; interorbital region relatively wide ..................................................................... 29
-Tympanic bullae swollen; alisphenoid canal absent; nasals usually elongate, and always wedged deeply between frontals; postcanine teeth usually bearing at least 3 cusps (may be obsolete or missing in some genera); postorbital processes absent; interorbital region narrowly constricted ............................................... Phocidae....... 33

29.  -Upper canines out of alignment with other teeth, extremely enlarged and ever-growing; upper incisor and postcanines all similar, and forming a continuous row; lower canine and postcanines all alike; dental formula usually I1/0 C1/1 P3/3 M0/0; mastoid process applied closely to auditory bulla and extending ventrally as a

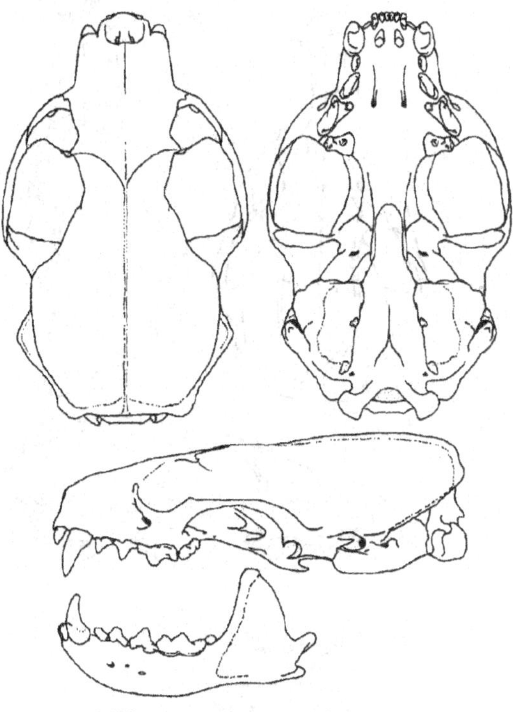

Fig. 121. *Mustela*
Greatest length of skull 65mm

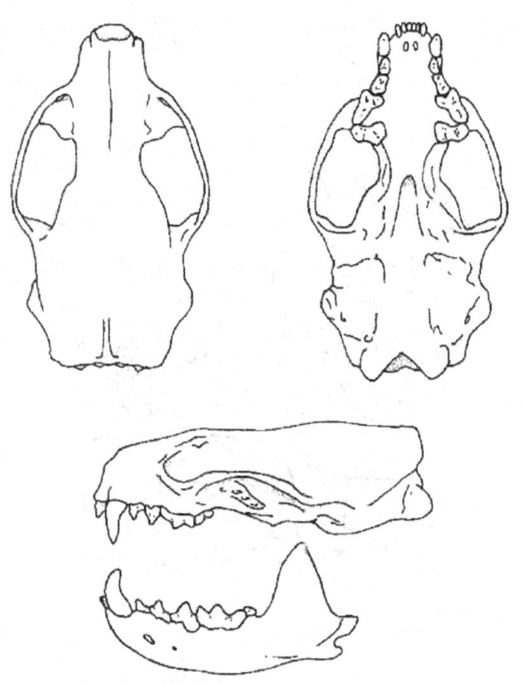

Fig. 122. *Neovison*
Greatest length of skull 69mm

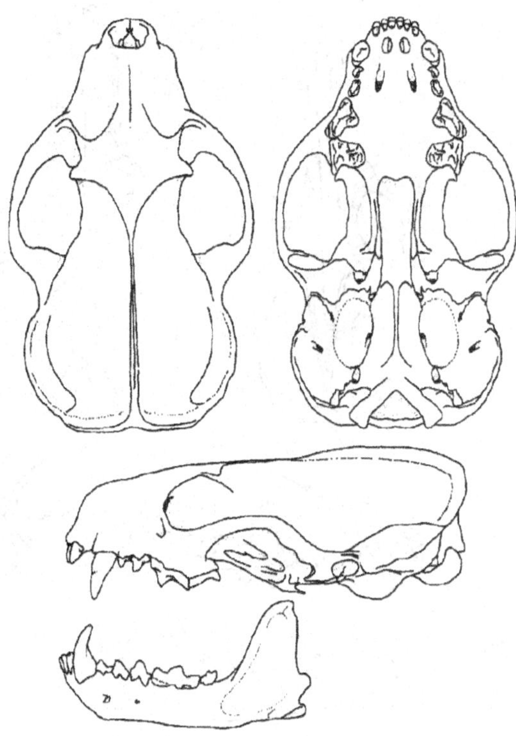

Fig. 123. *Spilogale*
Greatest length of skull 55mm

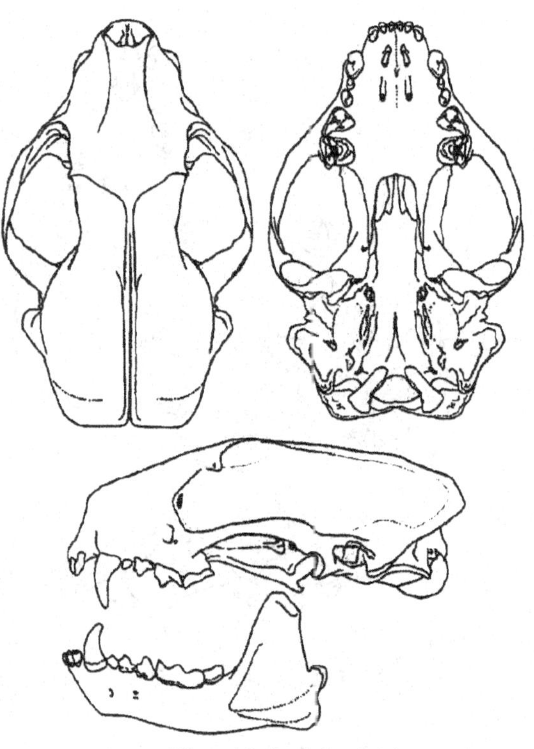

Fig. 124. *Mephitis*
Greatest length of skull 75mm

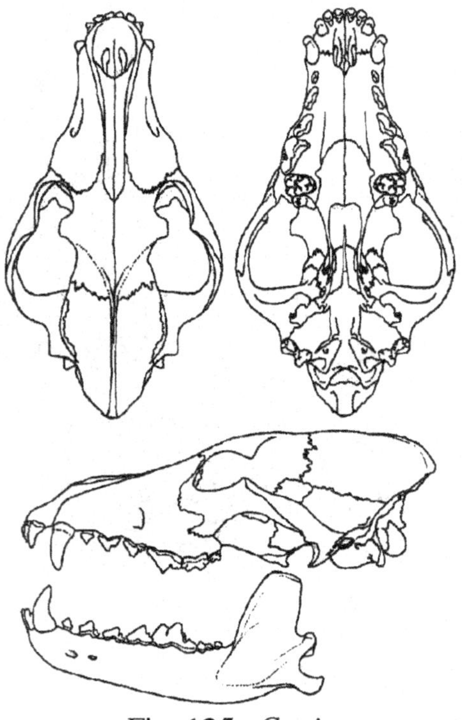

Fig. 125. *Canis*
Greatest length of skull 195mm

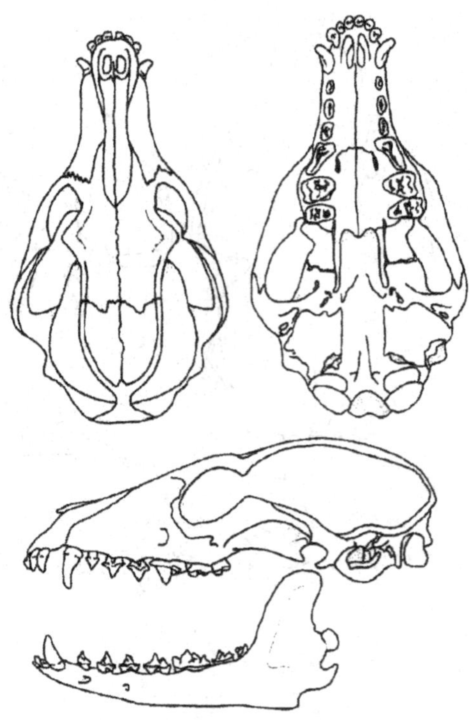

Fig. 126. *Urocyon*
Greatest length of skull 120mm

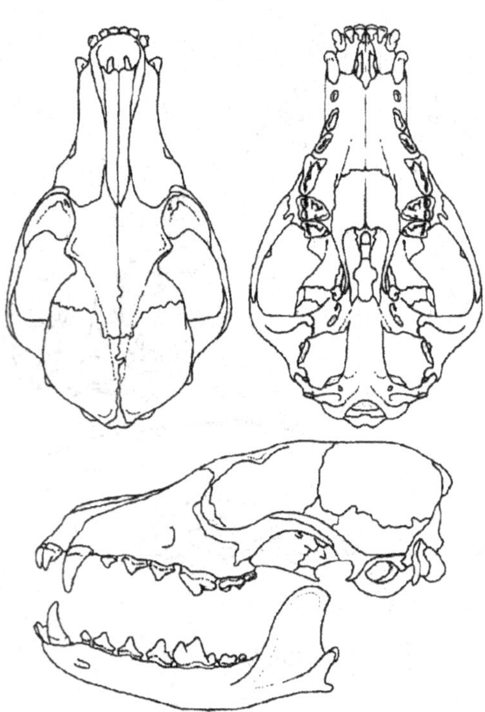

Fig. 127. *Alopex*
Greatest length of skull 125mm

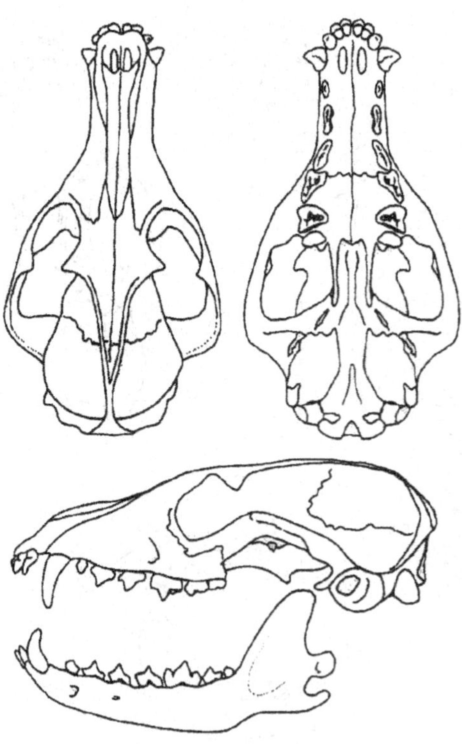

Fig. 128. *Vulpes*
Greatest length of skull 140mm

large projection; mandibular symphysis solidly fused; one mental foramen..................................................Odobenidae.. *Odobenus* (Fig. 129)
-Upper canines large but otherwise normal; dental formula I3/2 C1/1 P4/4 M1-2/1, $I^1$ and $I^2$ with transversely notched crowns, $I^3$ enlarged and caniniform; mastoid process not conspicuously enlarged, distinct from auditory bulla; mandibular symphysis not fused; 2 or more mental foramina................................Otariidae....... 30

30.  -Postorbital processes subquadrate; molar formula 1/1; a wide gap separating last upper postcanine from its predecessor..............................................................................*Eumetopias* (Fig. 130)
-Postorbital processes triangular; molar formula 2/1 or 1/1; upper postcanines uniformly spaced................................................................. 31

31.  -Upper toothrow usually composed of 5 teeth, each with a strong cingulum that often bears accessory cusplets....................*Zalophus* (Fig. 131)
-Upper toothrow usually composed of 6 teeth, cingulum absent or obsolete, no accessory cusps ................................................. 32

32.  -Postorbital processes directed posteriorly; greatest width across nasals almost equaling their length......................... *Callorhinus* (Fig. 132)
-Postorbital processes directed laterally; greatest width across nasals about one half their length................................. *Arctocephalus* (Fig. 133)

33.  -Upper incisors 2; premaxillae separated from nasals (except in *Monachus*)................................................................................ 34
-Upper incisors 3; premaxillae in contact with nasals................................................. 36

34.  -Incisors 2/2; premaxillae touching nasals ....................................*Monachus* (Fig. 134)
-Incisors 2/1; premaxillae not reaching nasals ........................................................... 35

35.  -Interpterygoid fossa narrower than palate; anterior border of bullae concave, interbullar space flat; median nasal septum not reaching front of nasals..................................... *Mirounga* (Fig. 135)
-Interpterygoid fossa fully as wide as palate; anterior border of bullae straight, interbullar space concave; nasal septum extending forward to tip of nasals..........................*Cystophora* (Fig. 136)

36.  -Jugal short and deep, its greatest height approximately equal to its greatest length; no sagittal crest ................................*Erignathus* (Fig. 137)
-Jugal considerably longer than high; sagittal crest usually present (absent in *Pusa*) ......................................................................... 37

37.  -Nasals horizontal, level with vertex of skull; superior border of rostrum straight from nasals to tip of

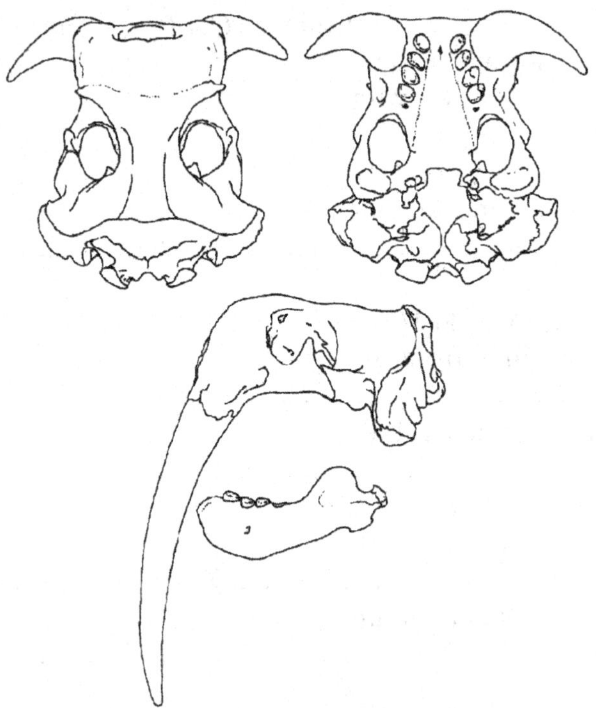

Fig. 129. *Odobenus*
Greatest length of skull 375mm

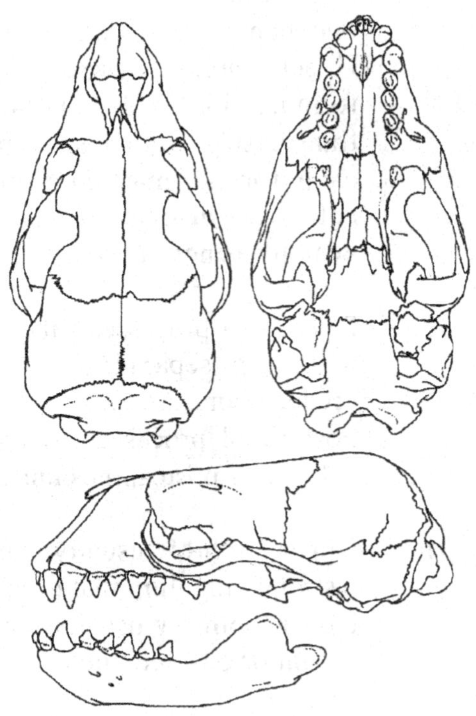

Fig. 130. *Eumetopias*
Greatest length of skull 325mm

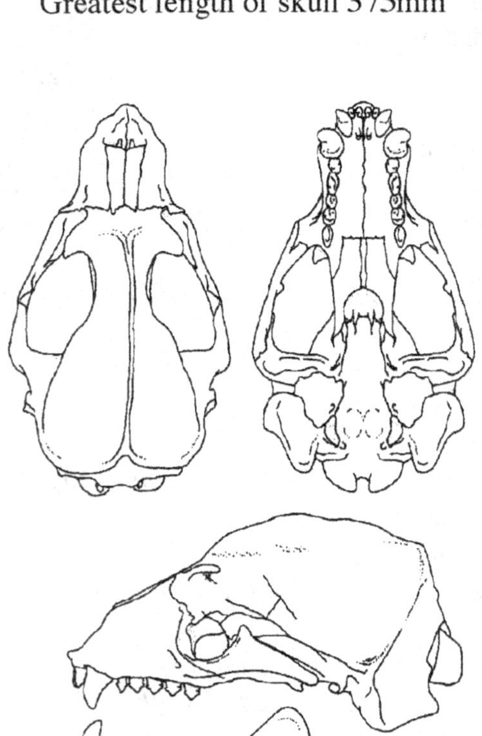

Fig. 131. *Zalophus*
Greatest length of skull 280mm

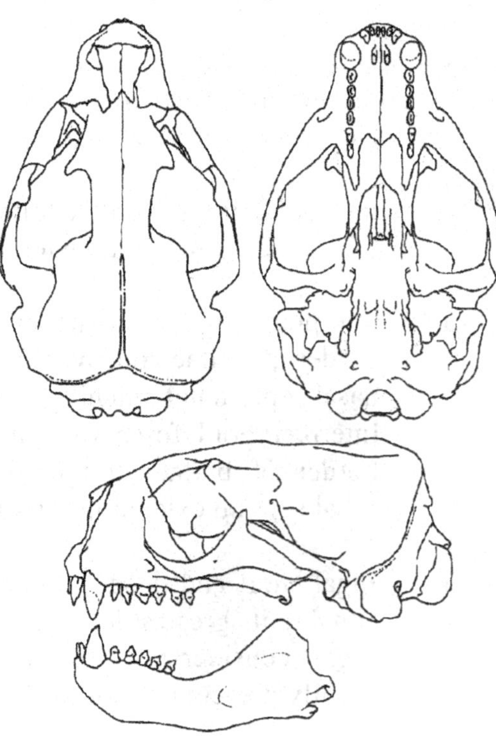

Fig. 132. *Callorhinus*
Greatest length of skull 225mm

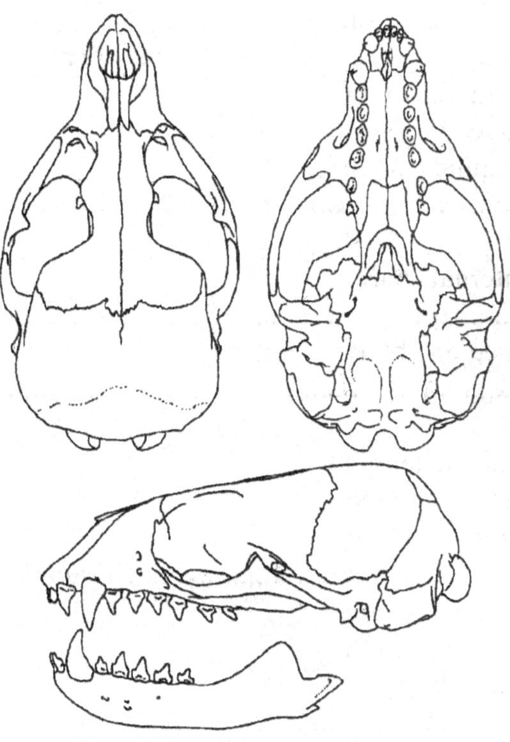

Fig. 133. *Arctocephalus*
Greatest length of skull 220mm

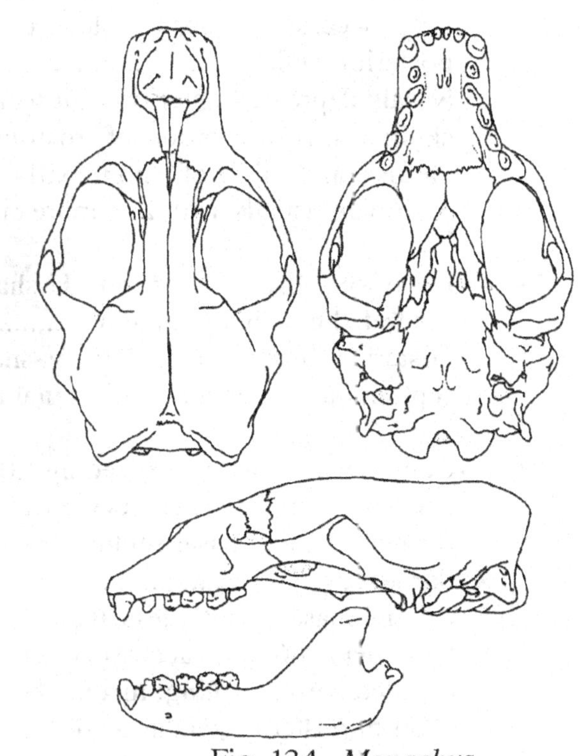

Fig. 134. *Monachus*
Greatest length of skull 285mm

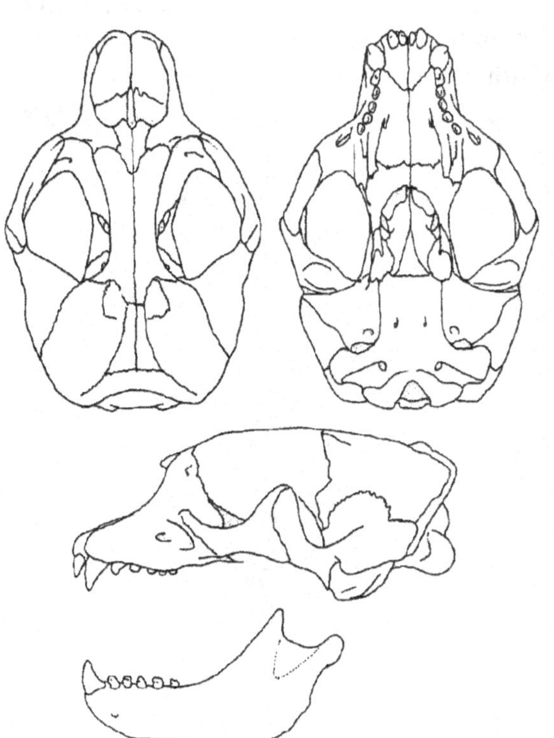

Fig. 135. *Mirounga*
Greatest length of skull 500mm

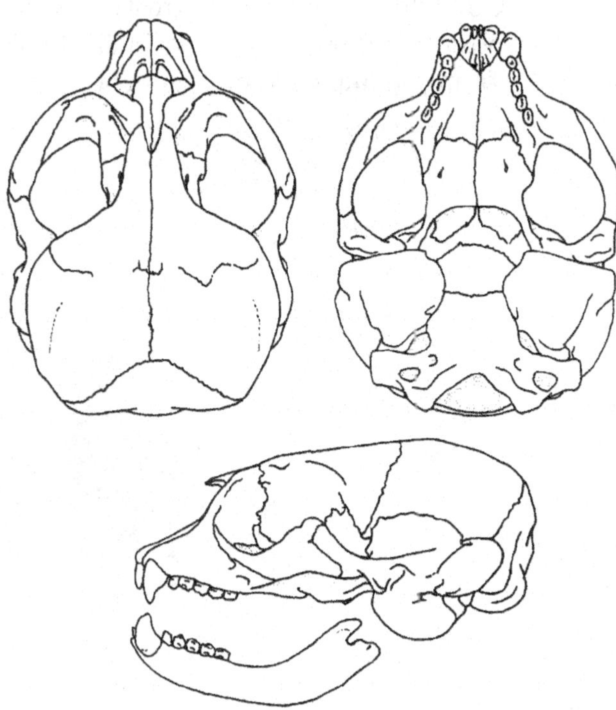

Fig. 136. *Cystophora*
Greatest length of skull 200mm

premaxillae; postcanine teeth robust, nearly as wide as long, accessory cusps obsolete, usually present only on posterior teeth....................................................................*Halichoerus* (Fig. 138)
-Nasals depressed anteriorly, descending from vertex of skull; superior border of rostrum concave between nasals and tip of premaxillae; postcanine teeth narrower, usually with 2 or more cusps each........................... *Phoca* (Fig. 139)........ 38

38.    -Posterior border of palate U-shaped; internal nares separated by a bony septum ............................................................................... 39
-Posterior border of palate V-shaped; median bony septum falling far short of internal nares ..................................................................... 40

39.    -Condylobasal length exceeding 200mm; palatal length exceeding 86mm; accessory cusps on cheek teeth well formed; median nasal septum reaching posterior palatal border ...............................................................................................*(Pagophilus)*
-Condylobasal length less than 200mm; palatal length less than 86mm; accessory cusps on cheek teeth obsolete, often lacking; median bony septum falling just short of posterior palatal border...............................................*(Histriophoca)*

40.    -Condylobasal length less than 187mm; postcanine teeth bearing only 3 cusps; interorbital width less than 7mm; sagittal crest lacking in adults.........................................................................*(Pusa)*
-Condylobasal length greater than 187mm; some postcanines with 4 cusps; interorbital width exceeding 7mm; sagittal crest present in adults.................................................................*(Phoca)*

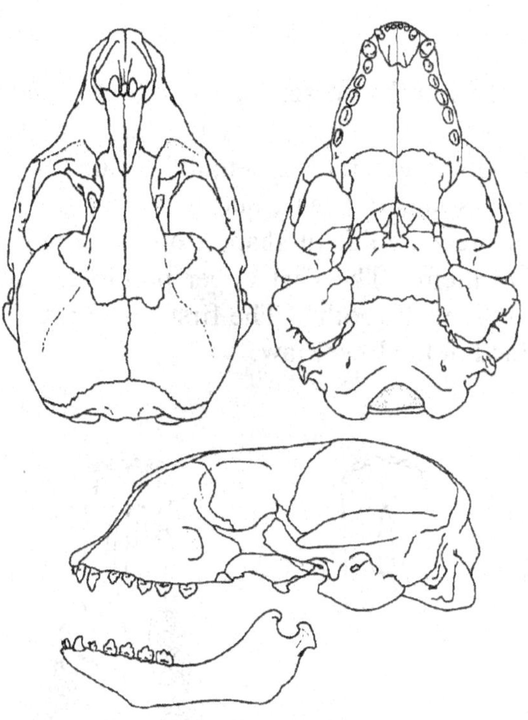

Fig. 137. *Erignathus*
Greatest length of skull 210mm

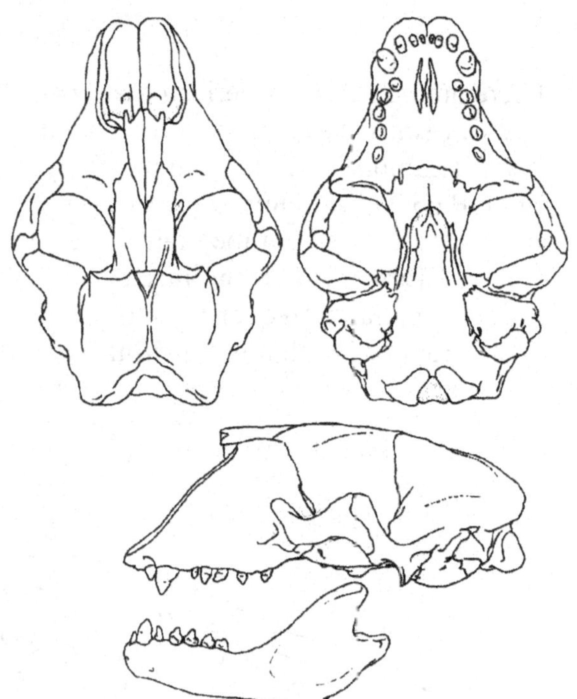

Fig. 138. *Halichoerus*
Greatest length of skull 290mm

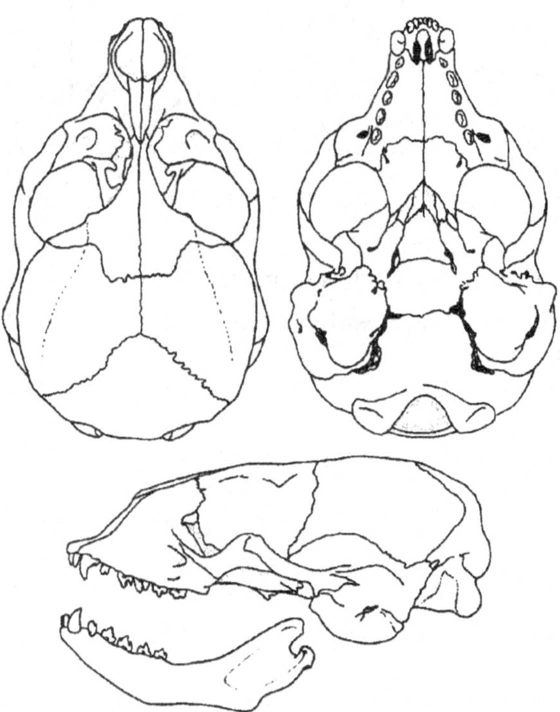

Fig. 139. *Phoca*
Greatest length of skull 170mm

There are no native American mammals of this order, but the horse, *Equus caballus*, and the ass, *E. asinus*, have been introduced and are living in many areas in the wild. The genus *Equus* (Equidae) may be recognized by the following characteristics: dental formula I3/3 C0-1/0-1 P3/3 M3/3 (canines are always present in jacks and stallions, but in mares, geldings, and mules they sometimes fail to develop); molars rectangular in shape; lower canines not crowded forward with the incisors and resembling them. The first upper premolar occurs with considerable frequency, and is referred to as the "wolf tooth". The first lower premolar rarely erupts, though it is sometimes found embedded in the lower jaw.

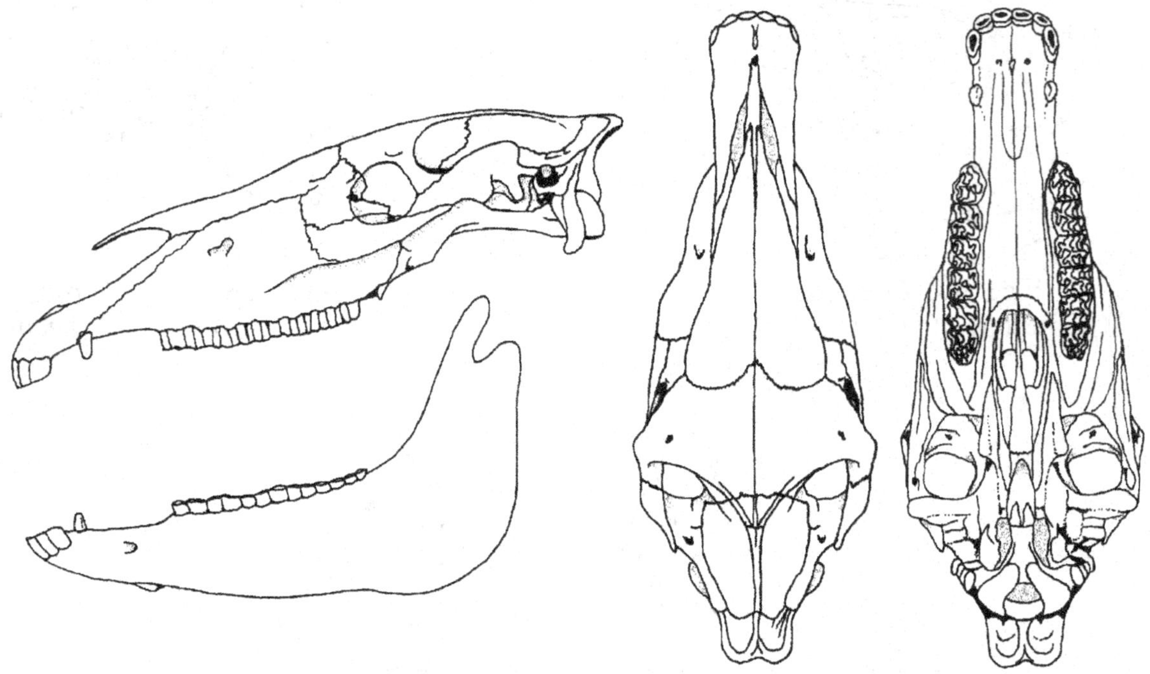

Fig. 140. *Equus**
Greatest length of skull 575mm

# ARTIODACTYLA

1.  -Incisors present above and below; upper and lower canines present; lower canines close to incisors but not modified to resemble them; cheek teeth bunodont or selenodont; postorbital bar incomplete to complete ............................................................. 2
    -Incisors absent above; upper canines present or absent, lower canines always present but situated close to the lower incisors and resembling them; cheek teeth selenodont; postorbital bar complete ................................................. 5

2.  -Incisors 1/3; upper and lower canines oval; postorbital bar complete ................................................................. Camelidae ......... 3
    -Incisors 2/3 or 3/3; upper and lower canines triangular; postorbital bar incomplete ........................................................ 4

3.  -Total length of skull greater than 350mm; dental formula I1/3 C1/1 P3/2 M3/3, total of 34 teeth ............................ *Camelus** (Fig. 141)
    -Total length of skull less than 350mm; dental formula I1/3 C1/1 P2/1 M3/3, total of 30 teeth ................................ *Lama** (Fig. 142)

4.  -Incisors 2/3, canines nearly straight, not sharply curved outward; in adults all sutures obliterated with age so that skull appears to be a single bone; crowns of jaw teeth slightly wrinkled; no free bone in nasal cartilage .............. Tayassuidae ................
    ........................................................................................ *Pecari* (Fig. 143)
    -Incisors 3/3, canines curved sharply outward; sutures of skull distinct throughout life; crowns of jaw teeth deeply wrinkled; free bone present in nasal cartilage ............. Suidae ..... *Sus** (Fig. 144)

5.  -Two lacrimal foramina on or just within anterior orbital rim; head ornamentation, when present, consisting of permanent bony cores surmounted by deciduous horny sheaths (Fig. 145) or deciduous bony antlers (Fig. 146) ................................. 6
    -One lacrimal foramen on or just within anterior orbital rim; head ornamentation, when present, consisting of permanent bony cores surmounted by permanent horny sheaths (Fig. 147) ................................................... Bovidae ....... 15

6.  -Antorbital pits lacking; ethmoid vacuities elongate and narrow, cancellous interior not broadly exposed; head ornamentation of permanent, erect, compressed, bony cores covered with a horny deciduous sheath; horns much larger with recurved points and an anterior prong in males, short and simple in females ............ Antilocapridae ..... *Antilocapra* (Fig. 145)

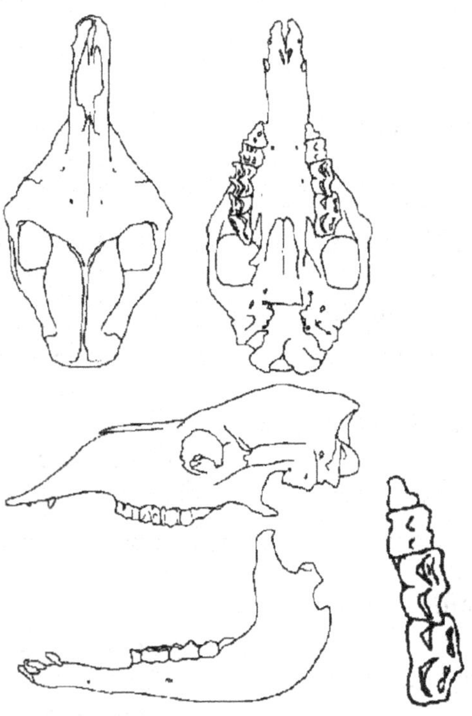

Fig. 141. *Camelus**
Greatest length of skull 470mm

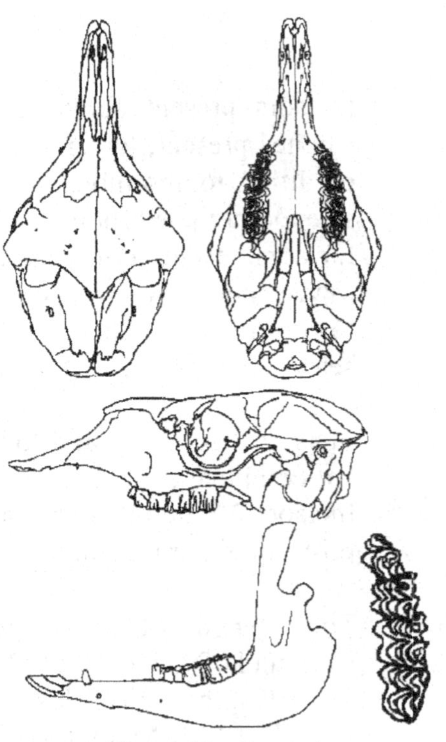

Fig. 142. *Lama**
Greatest length of skull 325mm

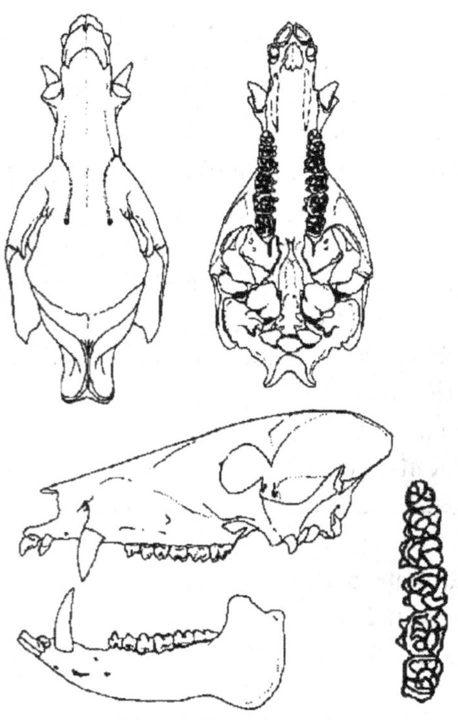

Fig. 143. *Pecari*
Greatest length of skull 250mm

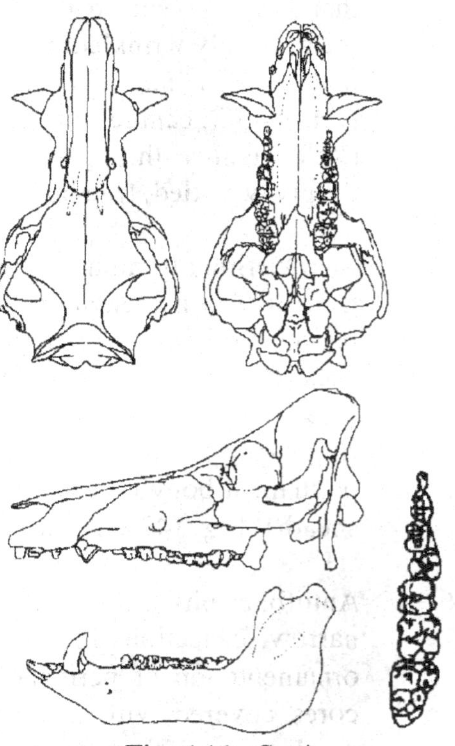

Fig. 144. *Sus**
Greatest length of skull 350mm

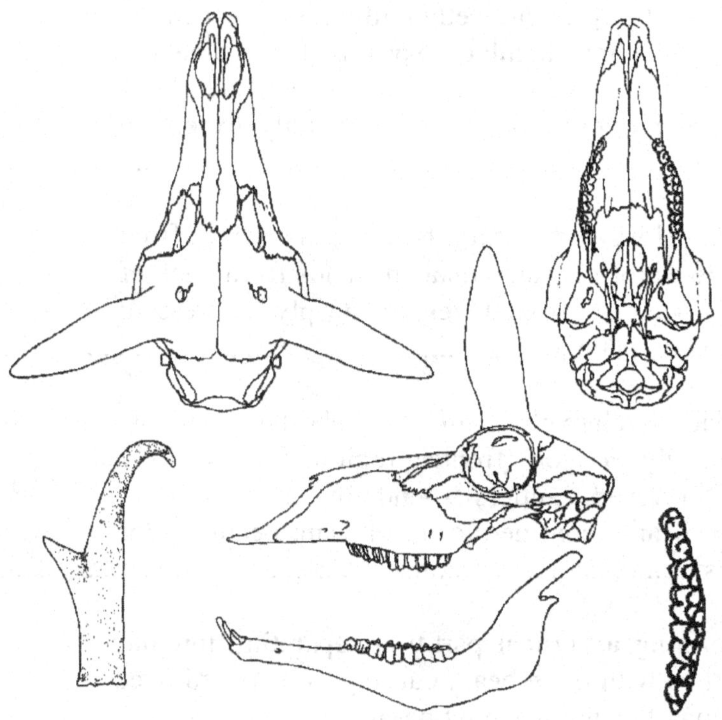

Fig. 145. *Antilocapra*
Greatest length of skull 250mm

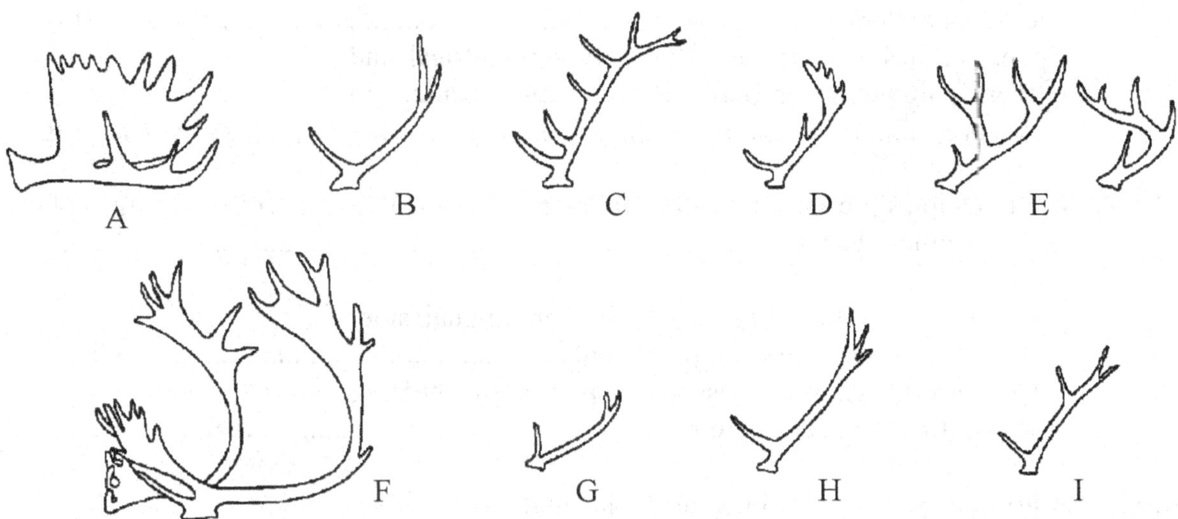

Fig. 146.  Antler types of A, *Alces*; B, *Axis**; C, (*Cervus*); D, *Dama**; E, *Odocoileus*; F, *Rangifer*; G, (*Rucervus*)*; H, (*Rusa*)*; and I, (*Sika*)*.  Note: Drawings are not to scale.

-Antorbital pits present; ethmoid vacuities large, triangular or subrectangular, exposing large areas of cancellous bone within the nasal cavity; head ornamentation bony antlers (in males only except in *Rangifer*) ..................................................... Cervidae......... 7

7.     -Premaxillae elongated, nasals short, distance from front of nasals to tip of rostrum equals distance from back of nasals to occiput; frontal region deeply depressed; antlers broadly palmate ...................................................*Alces* (Fig. 148)

       -Premaxillae relatively much shorter, nasals proportionately longer; frontal region flat or only slightly depressed; antlers round in cross section, palmation, if any, confined to tip of main beam and/or brow tines........................................................................ 8

8.     -Vomer forming a vertical partition separating internal nares; antlers with main beam curving far forward over head; premaxillae not reaching nasals ................................ 9

       -Vomer not dividing internal nares; antler with main beam directed posteriorly, not curving far forward over head; premaxillae in contact with nasals ........................ 10

9.     -Upper canines present; antlers asymmetrical, palmate at tip, and with one or more flattened tines descending over face; antlers present in both sexes.................... *Rangifer* (Fig. 149)

       -Upper canines absent; antlers usually symmetrical and without palmations or flattened brow tines; antlers in males only ................................................ *Odocoileus* (Fig. 150)

10.    -Upper canines present (not always bilateral in (*Rusa*)) ............. *Cervus* (Fig. 151).... 11

       -Upper canines absent ................................................................. 14

11.    -Upper molars with accessory cusplet on lingual side; antler bearing brow tine and royals only...............................................(*Rucervus*)*

       -Upper molars lacking accessory cusplet; antler bearing at least 2 tines in addition to royals........................................ 12

12.    -Antorbital pits encroaching on facial part of maxillae; antler 3-pointed, with brow, trez, and terminal point .....................................(*Rusa*)*

       -Antorbital pits entirely in lacrimals; antler with more than 3 points ........................................................................ 13

13.    -Size small, cranial length 200mm or less; antler with brow, trez, and 2 royals........................................................(*Sika*)*

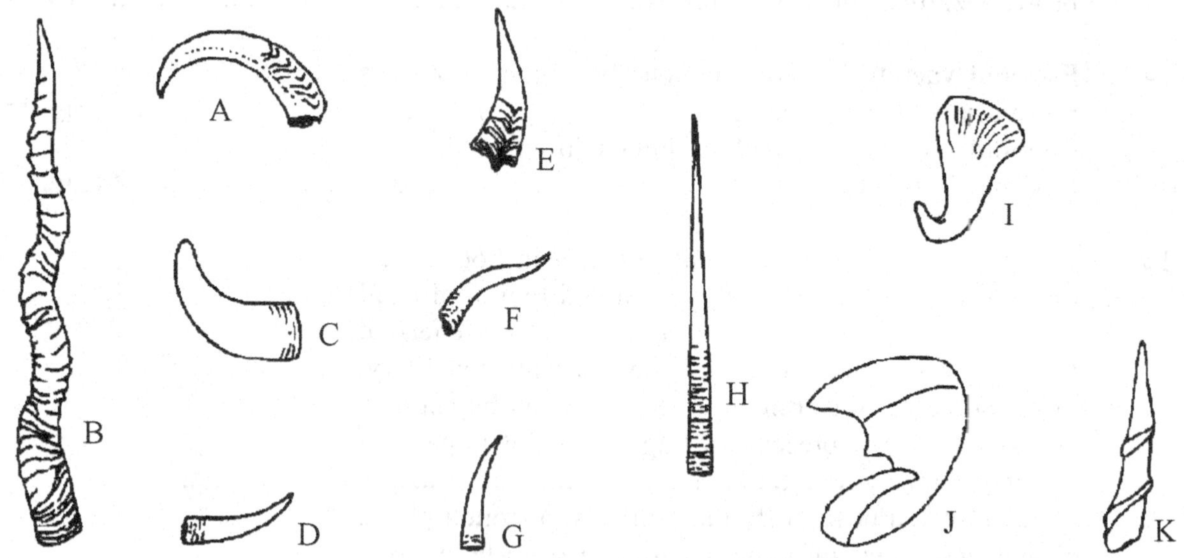

Fig. 147. Horn types of A, *Ammotragus**; B, *Antilope**; C, *Bison*; D, *Bos**; E, *Boselaphus**; F, *Capra**; G, *Oreamnos*; H, *Oryx**; I, *Ovibos*; J, *Ovis*; and K, *Taurotragus**. Note: Drawings are not to scale.

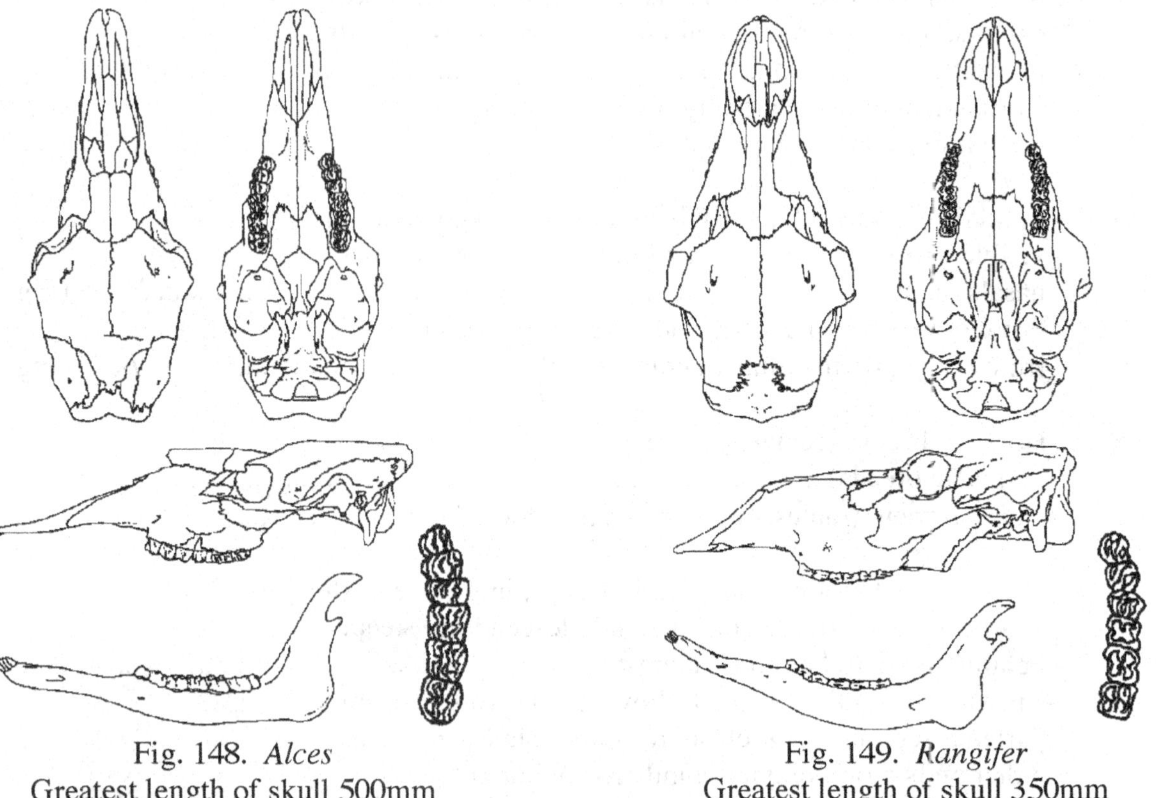

Fig. 148. *Alces*
Greatest length of skull 500mm

Fig. 149. *Rangifer*
Greatest length of skull 350mm

-Size large, cranial length 300mm or more; antler with
brow, bez, trez, and 2 or more royals.................................................... (*Cervus*)

14.    -Ethmoid vacuities subrectangular; brow and trez tines
present, royals palmated.................................................*Dama** (Fig. 152)
-Ethmoid vacuities triangular; antler with brow tine and
bifurcated tip only ..........................................................*Axis** (Fig. 153)

15.    -Horns directed laterally, usually present in both sexes,
(absent in some breeds of <u>Bos</u>); parietal region of skull
descending approximately at a right angle to forehead at
level of horns; ethmoid vacuities absent; maxillary,
lacrimal, nasal, and frontal bones in contact by sutures;
accessory cusplet present on lingual side of upper and
lower molars.................................................................................. 16
-Horns directed posteriorly more or less in frontal plane,
absent in females of some genera and from both sexes
in some breeds of *Ovis* and *Capra*; posterior portion of
cranium extending behind horns and forming an obtuse
angle with forehead; ethmoid vacuities present (except
in *Oreamnos*); no accessory cusplet on molars (except
*Oryx*) .......................................................................................... 18

16.    -Horns thick and massive, turning downward along side
of head behind orbits; basal bosses of horns
encroaching on top of cranium; shallow antorbital pits
present ..................................................................... *Ovibos* (Fig. 154)
-Horn cores directed laterally, usually curving upwards;
no massive basal bosses; antorbital pits absent.......................... 17

17.    -Frontals expanded laterally behind orbits; zygomata not
visible from above; premaxillae not in contact with
nasals...................................................................... *Bison* (Fig. 155)
-Frontals not expanded behind orbits; zygomata visible
from above; premaxillae in contact with nasals..................... *Bos** (Fig. 156)

18.    -Horns absent (females, and both sexes in some
domestic breeds) ........................................................................ 19
-Horns present (males, and females in some species) ................... 22

19.    -Crown elevated above and behind eyes; braincase with
nearly straight dorsal profile and descending steeply
behind crown to lambdoidal crest .............................................. 20
-Crown not highly elevated above orbits; forehead and
parietal region lying close to same plane; braincase
either arched or straight behind crown, but not

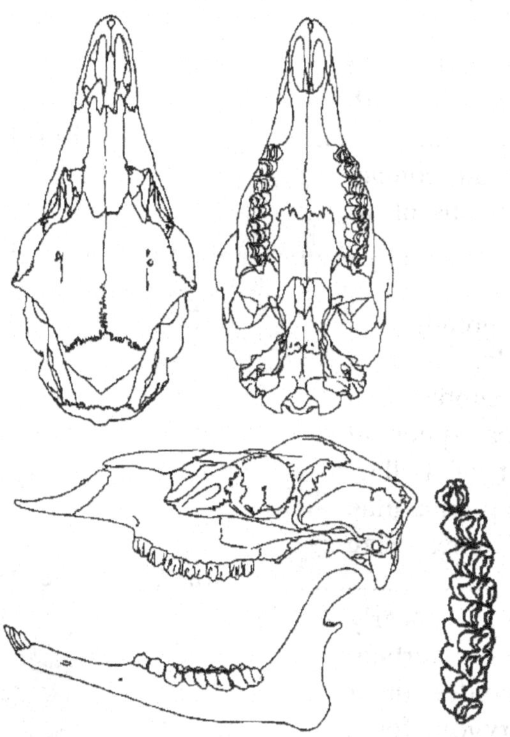

Fig. 150. *Odocoileus*
Greatest length of skull 270mm

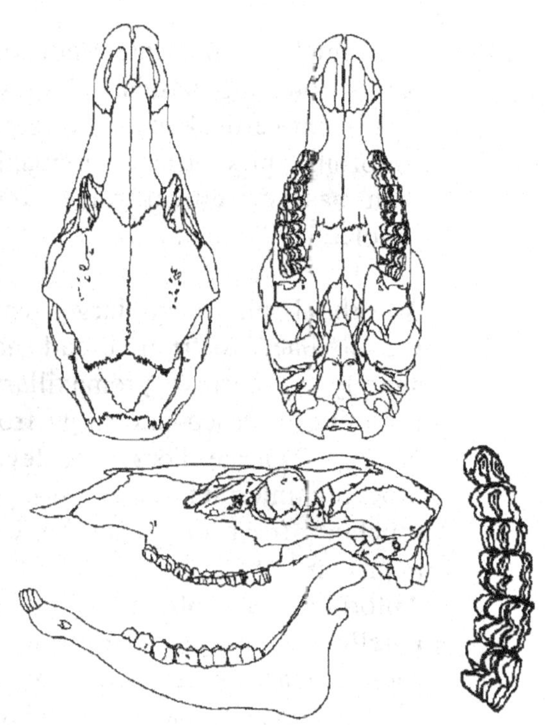

Fig. 151. *Cervus*
Greatest length of skull 425mm

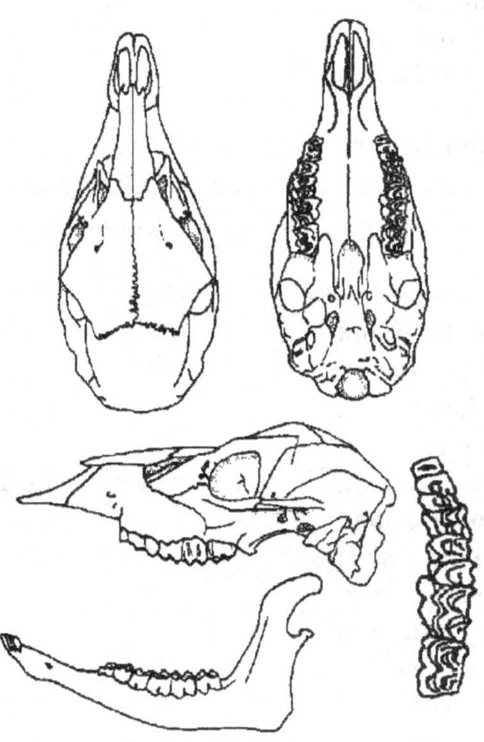

Fig. 152. *Dama**
Greatest length of skull 225mm

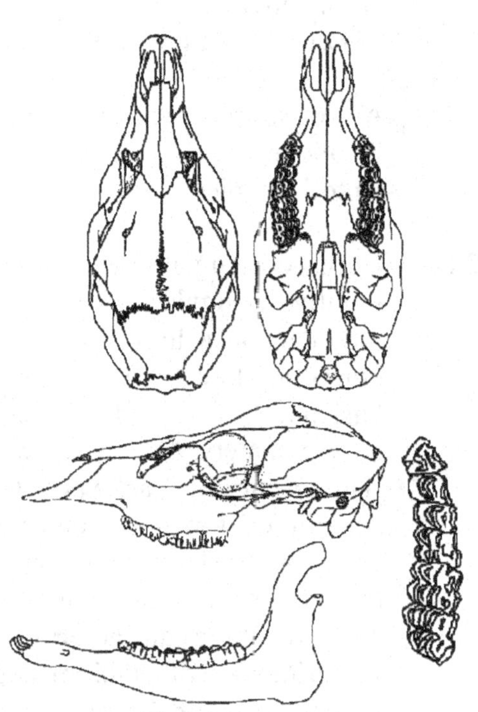

Fig. 153. *Axis**
Greatest length of skull 255mm

descending steeply to lambdoidal crest.............................................................. 21

20.    -Antorbital pits present; premaxillae falling short of or
barely touching nasals; labial side of premolars with
prominent vertical enamel ridges............................................... *Ovis* (Fig. 157)
-Antorbital pits absent; premaxillae in broad contact
with nasals; vertical ridges on premolars absent or
obsolete ............................................................................. *Capra** (Fig. 158)

21.    -Antorbital pits very large, encroaching broadly on
premaxillae; rostrum broad across nasals, tapering
sharply to narrow premaxillary region; supraorbital
foramina recessed into large frontal pits; basisphenoid
bearing 2 large bosses at level of front of bullae,
blocking interpterygoid fossa posteriorly; premaxillae
touching front of nasals; 5 lower postcanines; size
approximately 200mm in length .............................................*Antilope** (Fig. 159)
-Antorbital pits absent; sides of skull more nearly
parallel, tapering evenly towards front; supraorbital
foramina not recessed into pits; basisphenoid without
enlarged bosses and not blocking interpterygoid fossa
posteriorly; premaxillae reaching midpoint of nasals; 6
lower postcanines; size approximately 300mm in length ..........*Boselaphus** (Fig. 160)

22.    -Horns curling posteriorly, laterally, and downward................................... 23
-Horns directed posteriorly, straight, decurved, recurved,
or twisted, but not curling outward and downward ................................. 25

23.    -Antorbital pits present (for additional characteristics see
20 above)................................................................................ *Ovis* (Fig. 157)
-Antorbital pits absent ........................................................................ 24

24.    -Horns curving broadly outward and downward, cores
bluntly triangular in cross-section; base of horn not
mounted on a high bony pedicel; posterior portion of
cranium descending rather steeply; forehead flat;
basisphenoid and presphenoid meeting at
approximately 90° .................................................................*Ammotragus** (Fig. 161)
-Horns curving backward, then twisting outward, and
sometimes downward and inward in an inverse spiral;
horns with sharp crest on front edge; base of horn
mounted on a high pedicel; cross-section of horn core
pyriform with an anterior ridge; forehead concave;
posterior part of cranium depressed with basisphenoid
and presphenoid forming approximately a 30° angle to
plane of palate ..................................................................... *Capra** (Fig. 158)

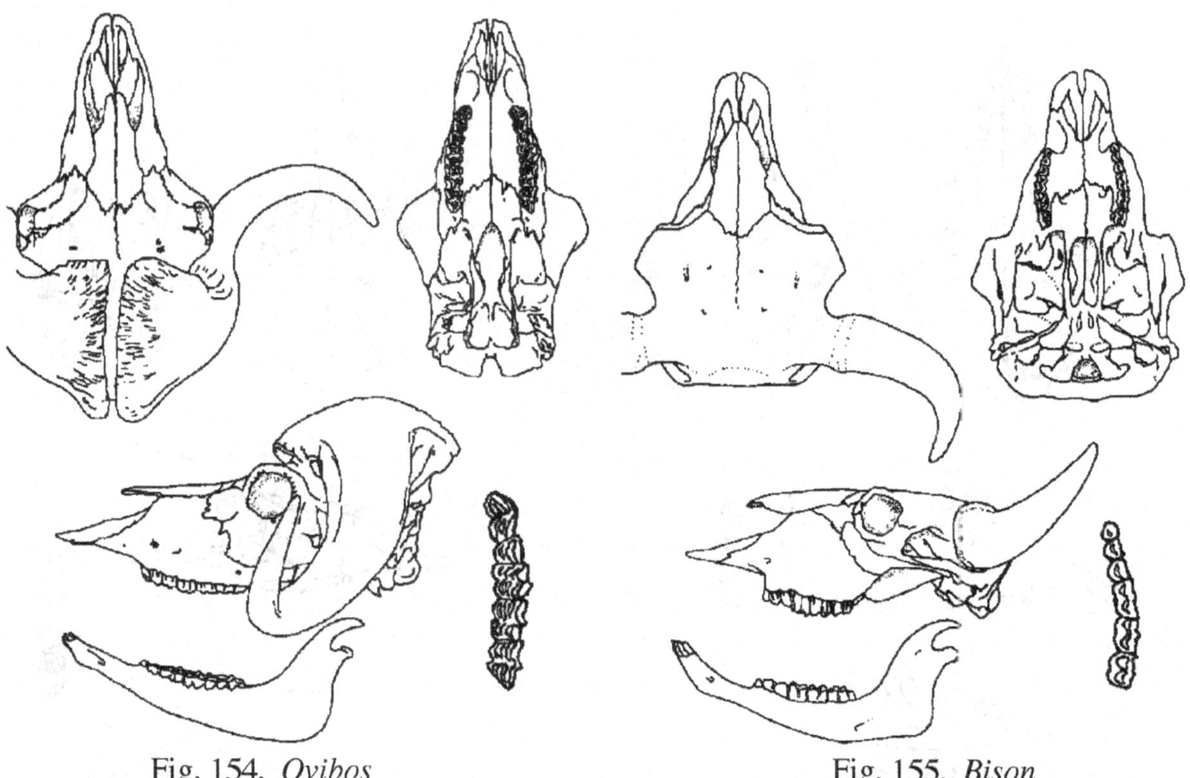

Fig. 154. *Ovibos*
Greatest length of skull 475mm

Fig. 155. *Bison*
Greatest length of skull 475mm

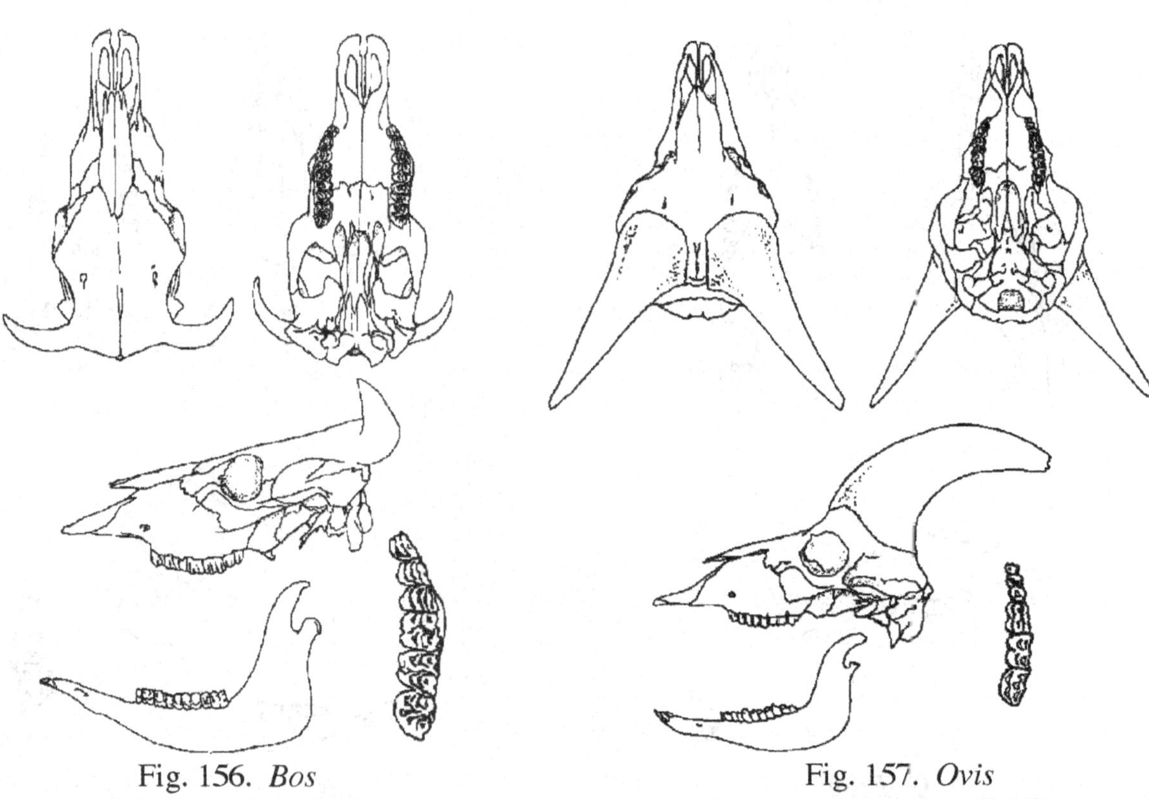

Fig. 156. *Bos*
Greatest length of skull 450mm

Fig. 157. *Ovis*
Greatest length of skull 250mm

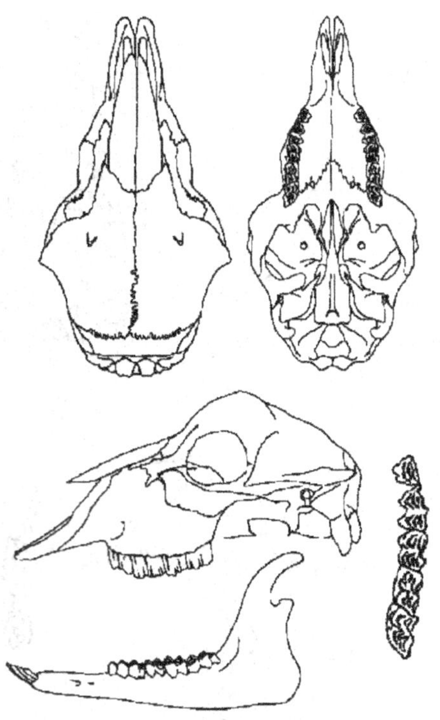

Fig. 158. *Capra**
Greatest length of skull 225mm

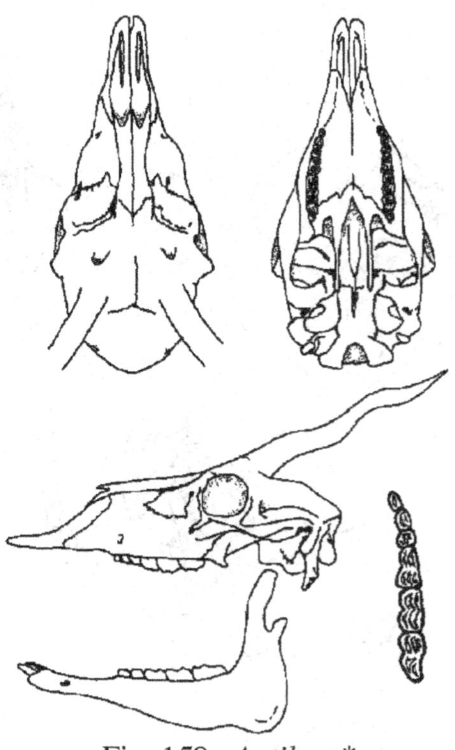

Fig. 159. *Antilope**
Greatest length of skull 200mm

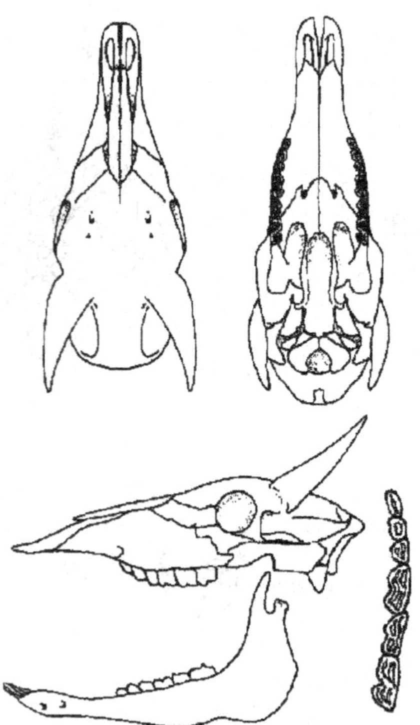

Fig. 160. *Boselaphus**
Greatest length of skull 425mm

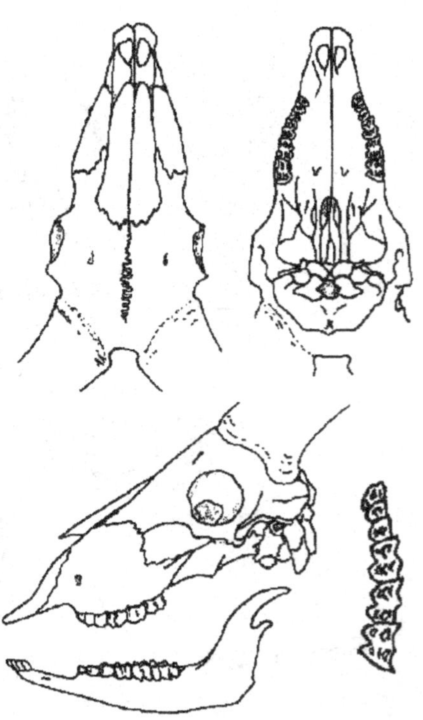

Fig. 161. *Ammotragus**
Greatest length of skull 275mm

25.     -Horn cores at least twice length of skull; antorbital pits
        present or absent .............................................................................. 26
        -Horn cores from less than to no more than 1.5 times
        length of skull; antorbital pits absent ........................................... 27

26.     -Horns and cores straight, ridged basally, and smooth for
        more than half their length; antorbital pits absent; skull
        250mm or more in length............................................ *Oryx** (Fig. 162)
        -Horns and cores corkscrew-spiraled, horns ridged
        nearly to tip; antorbital pits large, encroaching on
        maxillae (see 21 above for additional characters); skull
        approximately 250mm in length ............................*Antilope** (Fig. 159)

27.     -Horns and cores twisted in a tight spiral; supraorbital
        foramina recessed into a pair of deep frontal pits;
        ethmoid vacuities large; back of skull protruding
        downward well below level of palate; palate
        terminating opposite $M^2$; horns with a sharp ridge that
        encircles horn and crosses anterior surface twice ................... *Taurotragus** (Fig. 163)
        -Horn cores not twisted; supraorbital foramina not
        recessed into frontal pits; ethmoid vacuities absent or
        very small; back of skull not depressed below level of
        palate; palate terminating opposite or behind $M^3$ ....................... 28

28.     -Horns curving posteriorly; horn cores nearly straight,
        round in cross-section; ethmoid vacuities absent;
        premaxillae not contacting nasals .................................*Oreamnos* (Fig. 164)
        -Horns curving upward, bearing a distinct ridge on
        anterior surface of proximal part; horn cores curving
        upward and bearing an anterior ridge basally; ethmoid
        vacuities very narrow; premaxillae in contact with
        entire front half of nasals ........................................*Boselaphus** (Fig. 160)

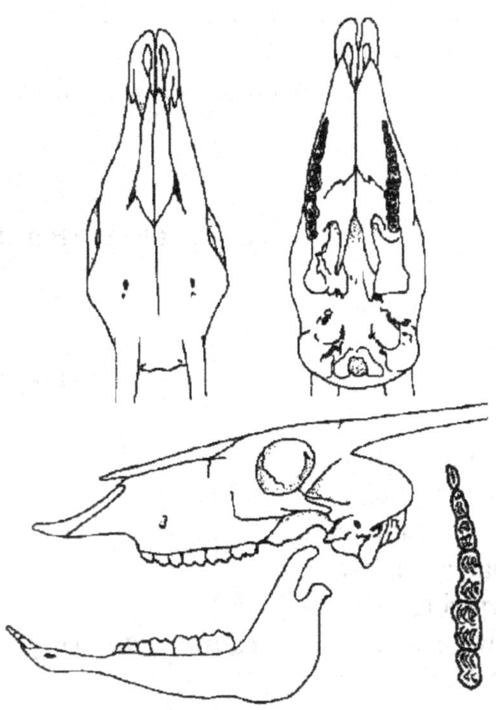

Fig. 162. *Oryx**
Greatest length of skull 300mm

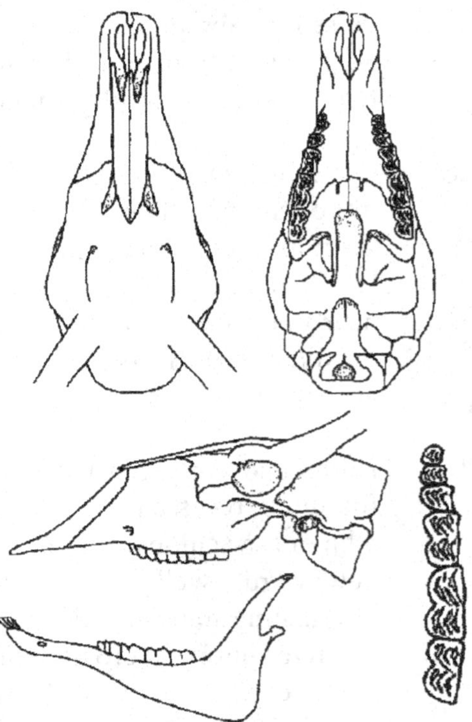

Fig. 163. *Taurotragus**
Greatest length of skull 485mm

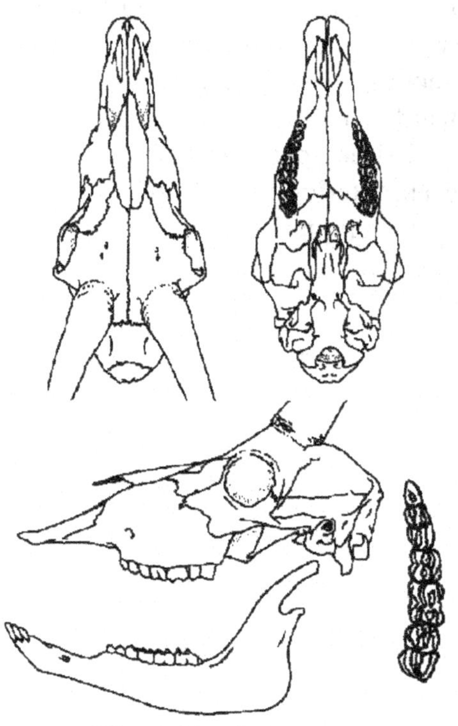

Fig. 164. *Oreamnos*
Greatest length of skull 275mm

# CETACEA

1.     -Teeth entirely absent; skull bilaterally symmetrical; lower jaw lacking a symphysis; occipital strongly inclined forward; nasals in front of vertex of skull, covering internal nares; size always very large, 250cm or more in length (see opposite choice for special case of *Physeter*) ..................................................................Suborder Mysticeti......... 2

        -Teeth present, sometimes only one or two pairs (may be embedded and invisible externally in *Monodon*); skull slightly to strongly asymmetrical; mandibles united to form a symphysis; occipital only slightly inclined forward; nasals on vertex of skull behind nares and not covering them; size considerably smaller, skull 150cm or less in length (usually much less) except in *Physeter*, in which case all other characteristics of this choice apply ............................................................ Suborder Odontoceti......... 7

2.     -Rostrum highly arched; each half of lower jaw strongly bowed outward; width across cranial region at least three times width of rostral base ...............................................Balaenidae......... 3

        -Skull, particularly rostrum, straight or moderately arched; rostrum extending forward like a beak; mandibles not strongly bowed outward; width across cranial region not more than twice width of rostral base............................................ 4

3.     -Rostrum relatively longer (condylobasal length about twice zygomatic width); in dorsal view zygomatic process of maxilla and supraorbital process of frontal directed obliquely backwards ........................................... *Balaena* (Fig. 165)

        -Rostrum relatively shorter (condylobasal length about 1.5 times zygomatic width); in dorsal view zygomatic process of maxilla and supraorbital process of frontal directed laterally perpendicular to median line of skull................ *Eubalaena* (Fig. 166)

4.     -Nasals large; frontals broadly exposed on vertex; anterior border of parietals behind posterior border of premaxillae, maxillae, and nasals; supraoccipital not extending forward beyond level of zygomatic process of squamosal; rostrum moderately arched .......Eschrichtidae.... *Eschrichtius* (Fig. 167)

        -Nasals reduced; frontals scarcely or not at all exposed on vertex; parietals extending forward laterally beyond level of posterior border of nasals and nasal processes of maxillae and premaxillae; supraoccipital extending forward beyond level of zygomatic process of squamosal; dorsal profile of rostrum nearly straight ................ Balaenopteridae......... 5

5.  -Cranial region approximately twice as wide as base of
    rostrum .........................................................................................*Megaptera* (Fig. 168)
    -Cranial region not more than 1.5 times as wide as base
    of rostrum....................................................................*Balaenoptera* (Fig. 169)......... 6

6.  -Borders of rostrum straight and evenly convergent from
    base to tip ....................................................................................................*(Balaenoptera)*

    -Borders of rostrum parallel proximally, and convergent
    towards tip....................................................................................................*(Sibbaldus)*

7.  -Rostral and anterior cranial region basin-like due to
    elevation of outer edges of maxillae; skull strongly
    asymmetrical, left nares many times larger than right;
    teeth confined to lower jaw (often one or two pairs in
    upper jaw of *Kogia*); mandible not extending to tip of
    rostrum (*Grampus* often has teeth in lower jaw only,
    but skull otherwise does not fit here)..................................... Physeteridae......... 8
    -Rostral and anterior cranial region not extensively
    basinlike, or if so, basin is enclosed by elevations of
    premaxillae only (*Ziphius*); skull slightly to strongly
    asymmetrical; bony nares not greatly different in size;
    teeth usually numerous in both jaws (except in
    *Grampus*, *Monodon* and the Ziphiidae); lower jaw
    extending approximately to tip of rostrum................................................. 9

8.  -Rostrum very short and wedge-shaped, only 0.3 of total
    skull length; lower jaw with 8-9 pairs of teeth;
    mandibular symphysis short; jugal absent, zygomata
    incomplete...................................................................................*Kogia* (Fig. 170)
    -Rostrum proportionately long, about 0.6 of total skull
    length; lower jaw with 20 or more pairs of teeth;
    mandibular symphysis joined for more than 1/2 length
    of mandible; jugal present, zygomata complete ............................. *Physeter* (Fig. 171)

9.  -Enlarged teeth restricted to lower jaw, but upper jaw
    may bear many vestigial teeth; mandible bearing only
    one or two pairs of large functional teeth; if two, then
    anterior much larger than posterior pair; posterior
    extensions of maxillary bones behind nasals elevated to
    form conspicuous asymmetrical crests; rostrum
    elongate, slender, and tapering, and deeper than wide
    (only as deep as wide in *Berardius*)..................................................Ziphiidae....... 10

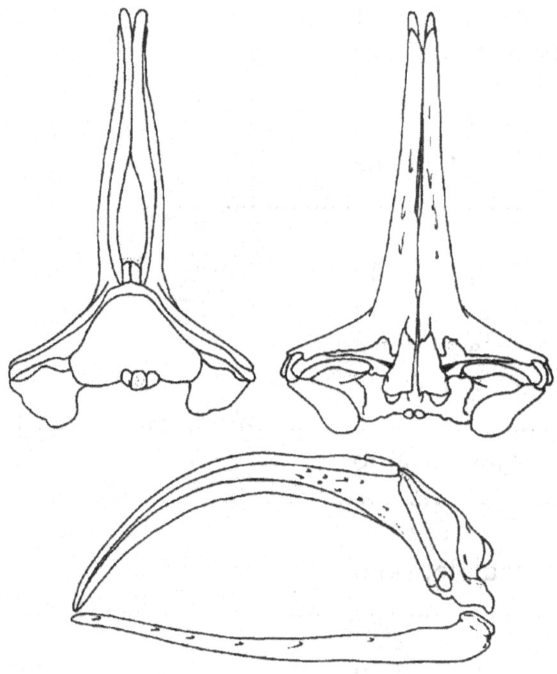

Fig. 165. *Balaena*
Greatest length of skull 4800mm

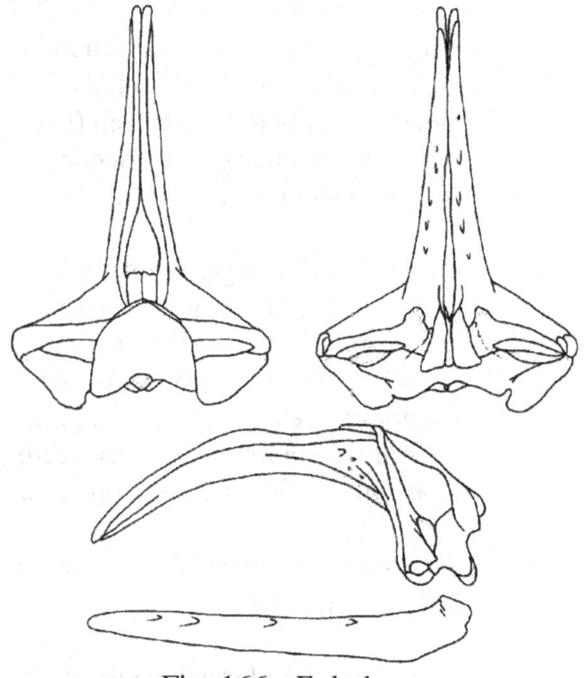

Fig. 166. *Eubalaena*
Greatest length of skull 3100mm

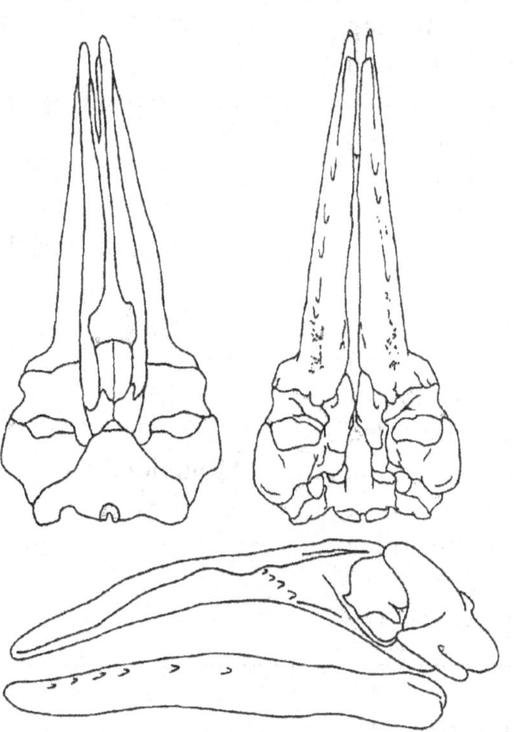

Fig. 167. *Eschrichtius*
Greatest length of skull 2450mm

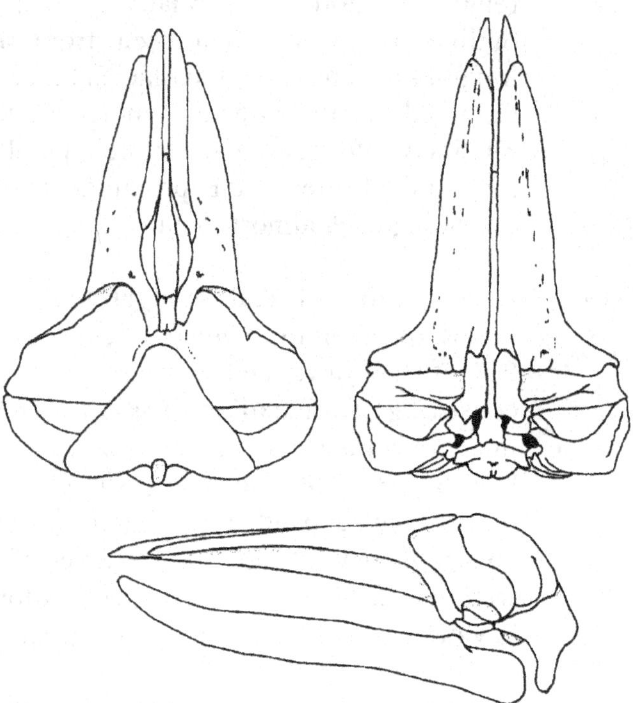

Fig. 168. *Megaptera*
Greatest length of skull 2775mm

-Enlarged teeth usually in both jaws (*Grampus* may have lower teeth only, but these are 2-7 pairs, and all are of equivalent size); no conspicuous crests behind or above nasal orifice; rostrum relatively blunt, not long and slender, and wider than deep (except in *Monodon* where one side is enlarged to accommodate the root of the tusk in males) ....................................................................... 13

10.  -One pair of enlarged teeth in lower jaw; teeth laterally compressed, the long dimension at least twice the short dimension; functional teeth situated some distance from tip of mandible, behind posterior end of symphysis except in one species.................................*Mesoplodon* (Fig. 172)
     -One or two pairs of enlarged teeth in lower jaw, oval to nearly round in cross-section ............................................. 11

11.  -Two pairs of enlarged teeth near tip of mandible; skull nearly symmetrical..................................................... *Berardius* (Fig. 173)

     -One pair of enlarged teeth at tip of lower jaw; skull asymmetrical .......................................................................... 12

12.  -Enlarged teeth directed forward and upward; rostrum relatively short, with relatively wide base, making outline triangular when seen from above; no high longitudinal crests on maxillae ......................................... *Ziphius* (Fig. 174)
     -Enlarged teeth directed more vertically; rostrum relatively longer and more parallel-sided; high longitudinal crests on proximal parts of maxillae, extending back almost to nares ..................................... *Hyperoodon* (Fig. 175)

13.  -Dorsal profile of skull straight from tip of rostrum to vertex of cranium; teeth few, either restricted to front 2/3 of the jaws and pointing strongly forward, or consisting of a maxillary tusk on each side, embedded in the maxilla except in males.............................. Monodontidae....... 14
     -Dorsal profile of skull rising sharply at level of nares, nasals and expanded proximal part of premaxillae conspicuously elevated above level of top of rostrum; teeth usually numerous in upper and lower jaws (upper jaw often toothless in *Grampus*) ....................................... 15

14.  -Teeth small, pointing forward and confined to front 2/3 of jaws.............................................................*Delphinapterus* (Fig. 176)
     -A forward-pointing spiral tusk usually erupting from anterior end of left (occasionally right) maxilla in

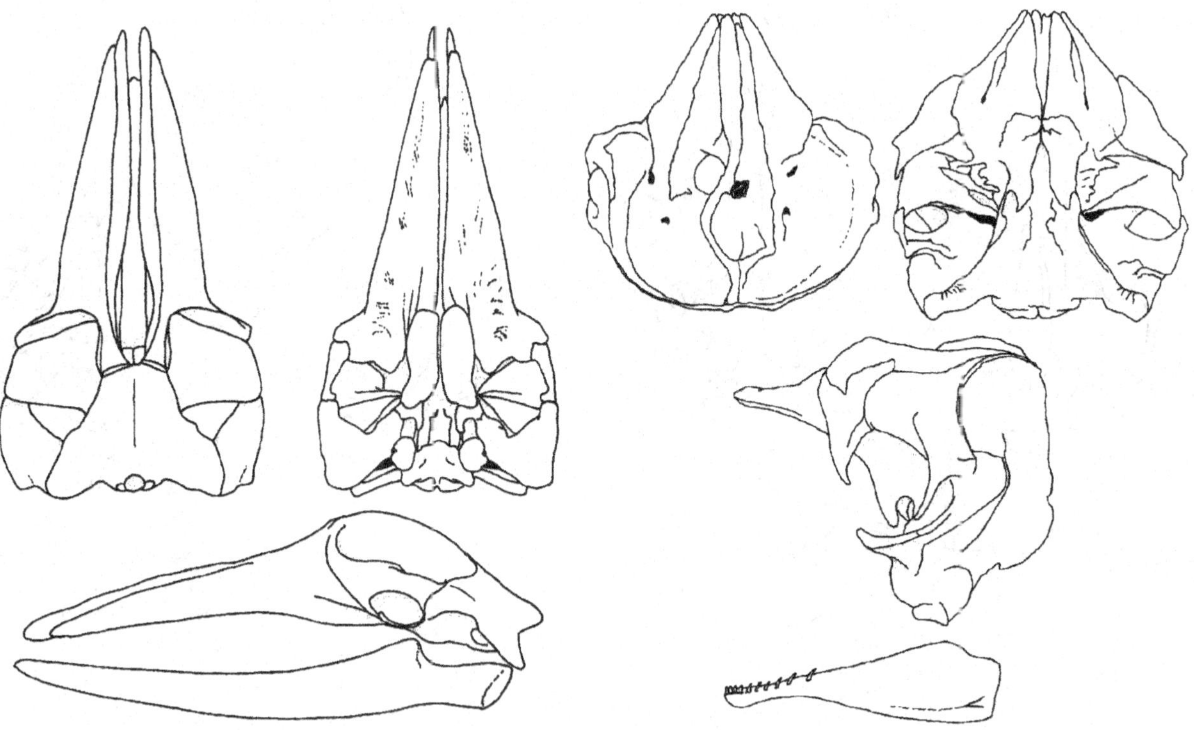

Fig. 169. *Balaenoptera*
Greatest length of skull 1100mm

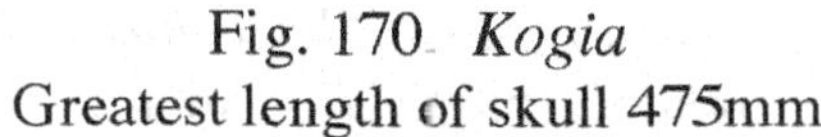

Fig. 170. *Kogia*
Greatest length of skull 475mm

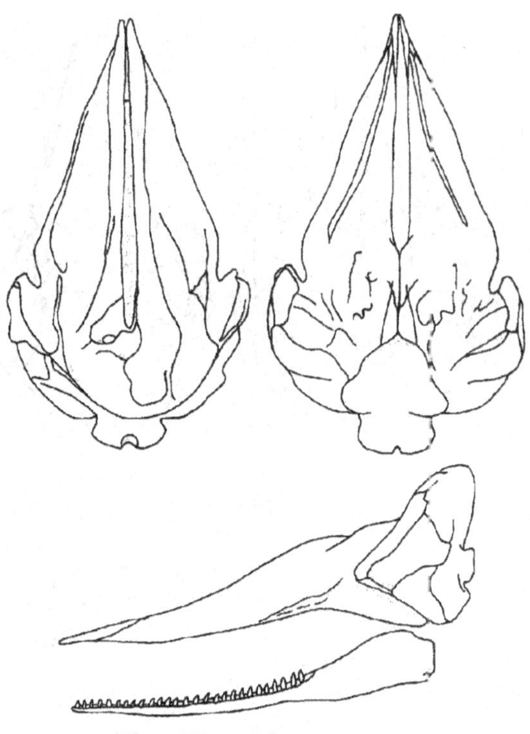

Fig. 171. *Physeter*
Greatest length of skull 3150mm

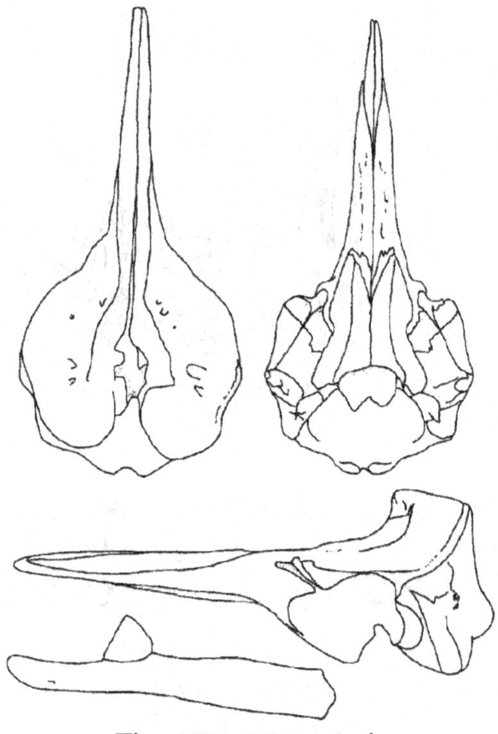

Fig. 172. *Mesoplodon*
Greatest length of skull 670mm

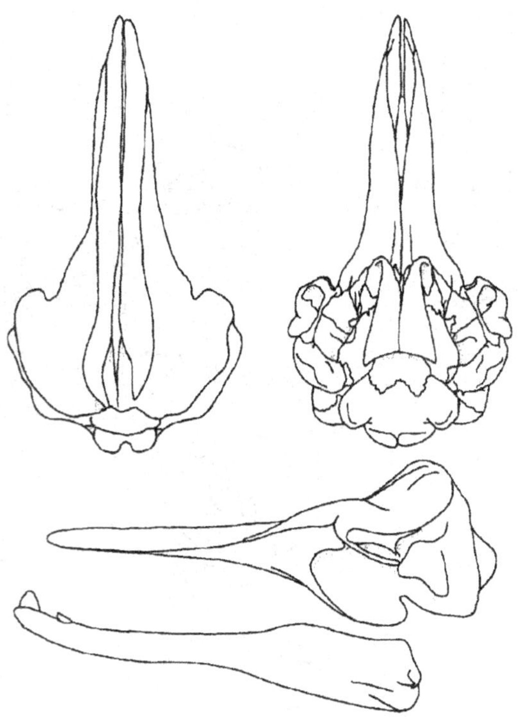

Fig. 173. *Berardius*
Greatest length of skull 1400mm

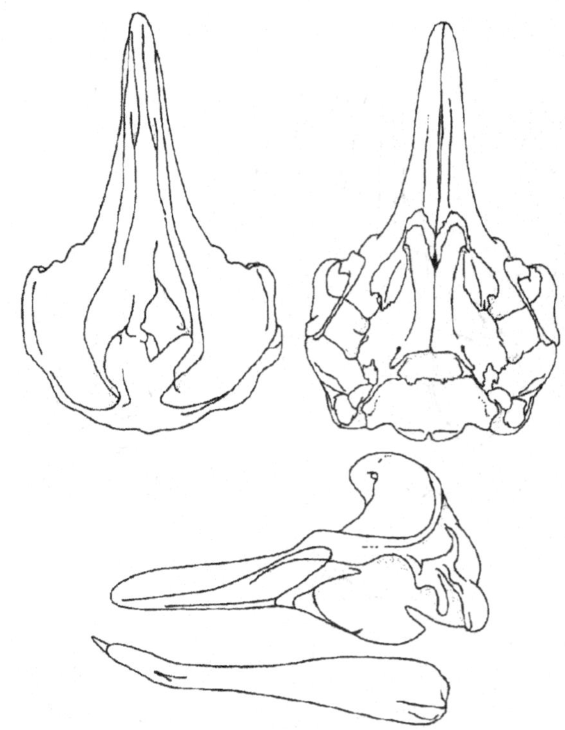

Fig. 174. *Ziphius*
Greatest length of skull 900mm

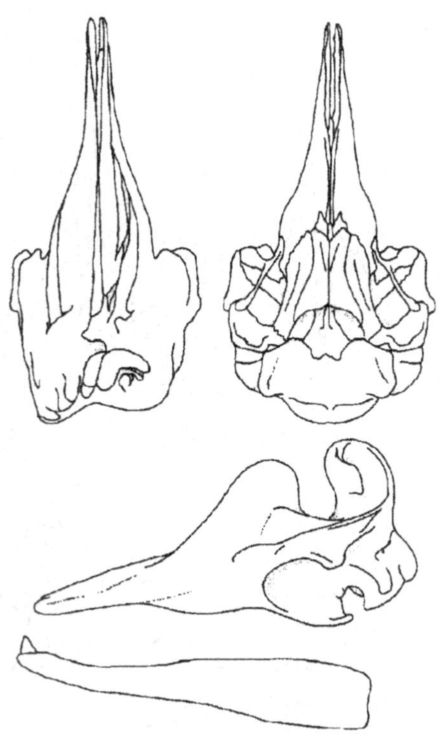

Fig. 175. *Hyperoodon*
Greatest length of skull 1050mm

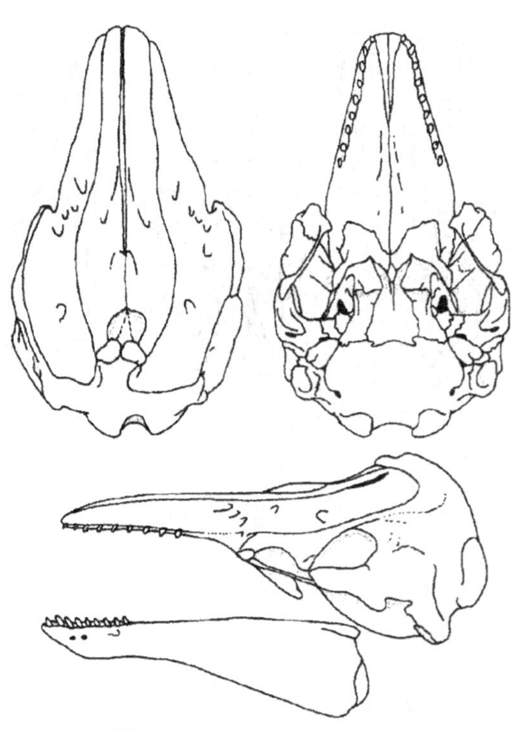

Fig. 176. *Delphinapterus*
Greatest length of skull 425mm

males; right tusk present, but embedded in bone, as are
both tusks in females, which appear toothless ................................ *Monodon* (Fig. 177)

15.    -Teeth spade-shaped in lateral view; proximal ends of
premaxillae with raised bosses in front of nares ........................... Phocoenidae ....... 16
-Teeth circular or oval in cross-section and not spade-
shaped; proximal ends of premaxillae flat or shallowly
depressed ................................................................................... Delphinidae ....... 17

16.    -Teeth normal in size and fully erupted ......................................... *Phocoena* (Fig. 178)
-Teeth much reduced, often so small as not to erupt
through the gums ........................................................................ *Phocoenoides* (Fig. 179)

17.    -Teeth with rough-surfaced crowns; symphysis of
mandible 1/4-1/3 of total mandibular length ............................... *Steno* (Fig. 180)
-Teeth with smooth crowns; mandibular symphysis less
than 1/5 of total mandibular length ........................................................ 18

18.    -Teeth 2-7 pairs and confined to symphyseal region of
lower jaw (some individuals have 1-2 pairs in upper
jaw) ............................................................................... *Grampus* (Fig. 181)
-Teeth 7 or more pairs in both upper and lower tooth
rows .............................................................................................. 19

19.    -Teeth 13 or fewer pairs in both upper and lower rows ............................... 20
-Teeth 19 or more pairs in both upper and lower rows ............................... 23

20.    -Teeth small, confined to anterior half of rostrum ..................... *Globicephala* (Fig. 182)
-Teeth large and robust, occupying more than anterior
half of rostrum ....................................................................... 21

21.    -Pterygoids not in contact at posterior midline of palate;
teeth oval in cross-section ....................................................... *Orcinus* (Fig. 183)
-Pterygoids touching; teeth circular in cross-section ............................... 22

22.    -Total length of adult skull exceeding 500mm .............................. *Pseudorca* (Fig. 184)
-Total length of adult skull less than 500mm ............................. *Feresa* (Fig. 185)

23.    -Palate with two deep longitudinal grooves ................................... *Delphinus* (Fig. 186)
-Palate smooth or with only very shallow grooves ............................... 24

24.    -Premaxillae separated along entire length of rostrum ........... *Peponocephala* (Fig. 187)
-Premaxillae convergent and in contact at least along
distal half of rostrum, often actually fused ............................... 25

25.    -Rostral portion of premaxillae convex in cross-section ..................... *Stenella* (Fig. 188)

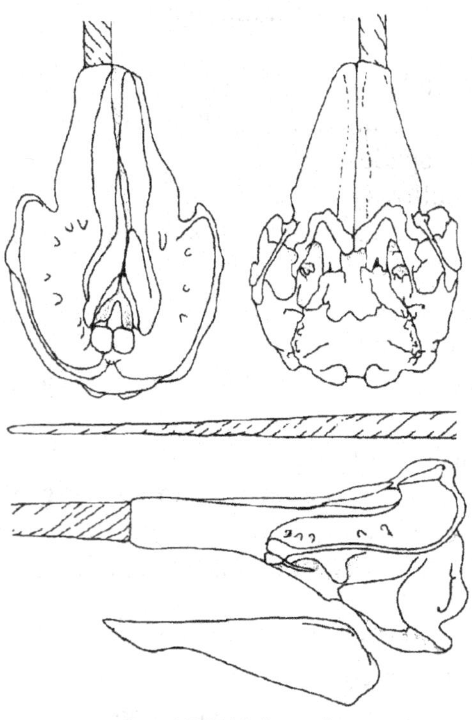

Fig. 177. *Monodon*
Greatest length of skull 1900mm

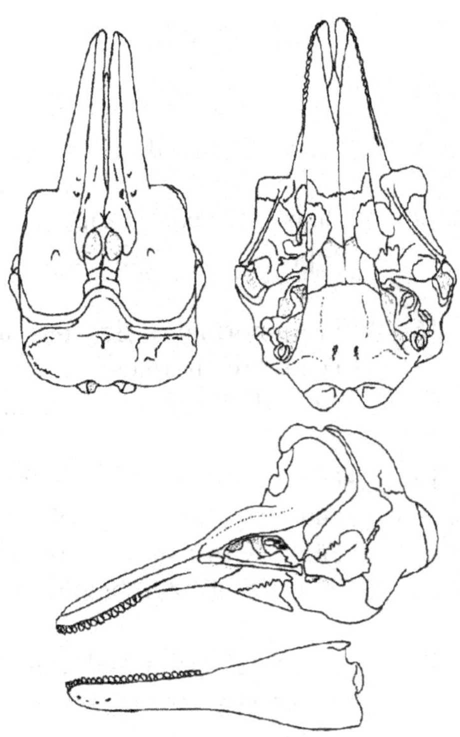

Fig. 178. *Phocoena*
Greatest length of skull 315mm

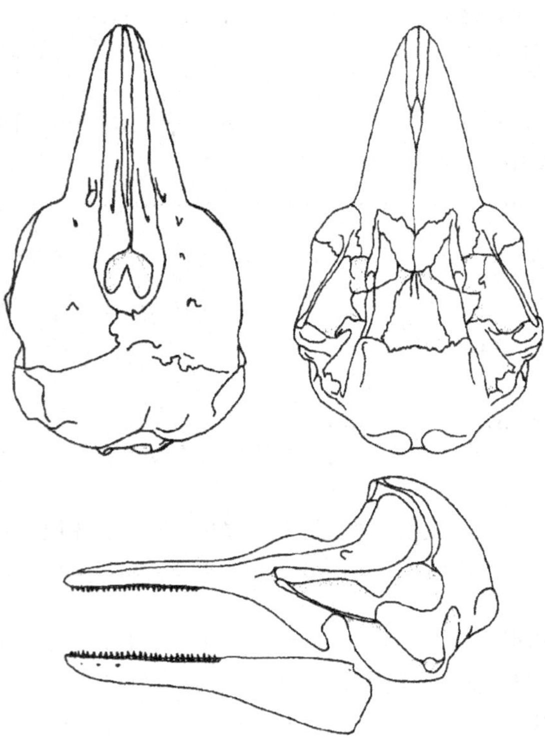

Fig. 179. *Phocoenoides*
Greatest length of skull 330mm

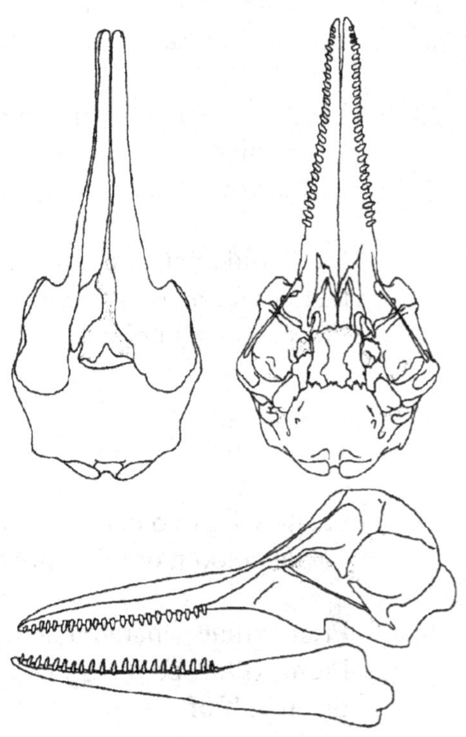

Fig. 180. *Steno*
Greatest length of skull 540mm

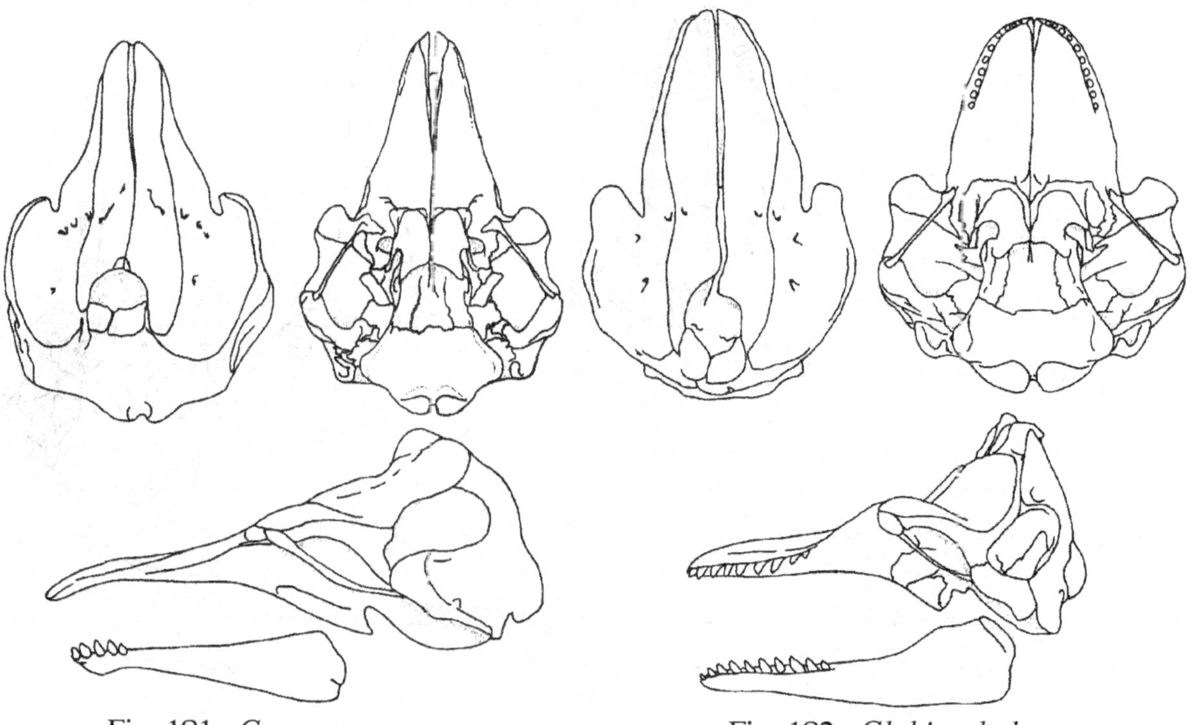

Fig. 181. *Grampus*
Greatest length of skull 475mm

Fig. 182. *Globicephala*
Greatest length of skull 700mm

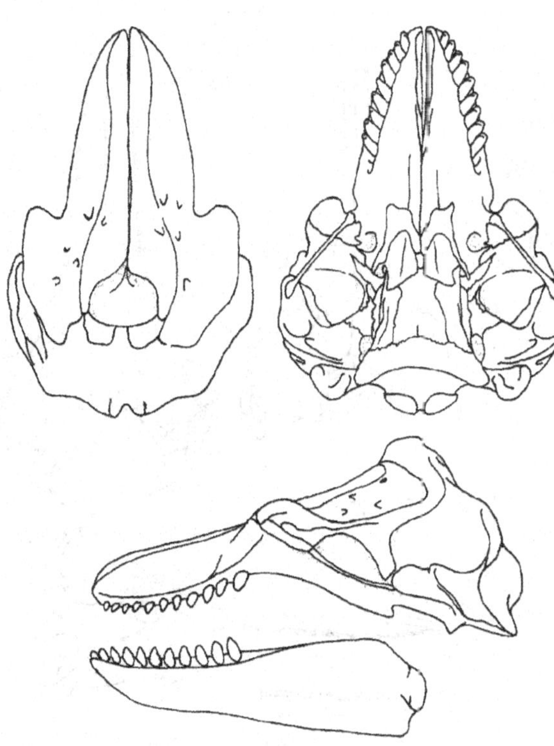

Fig. 183. *Orcinus*
Greatest length of skull 1225mm

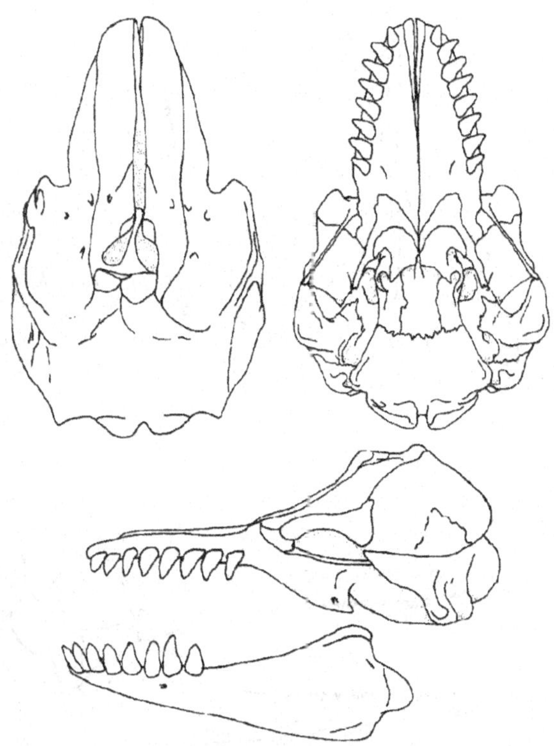

Fig. 184. *Pseudorca*
Greatest length of skull 525mm

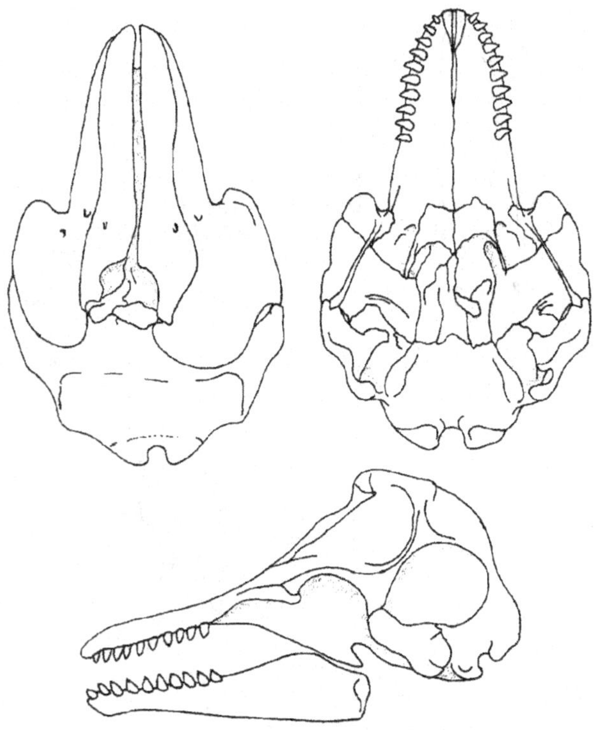

Fig. 185. *Feresa*
Greatest length of skull 425mm

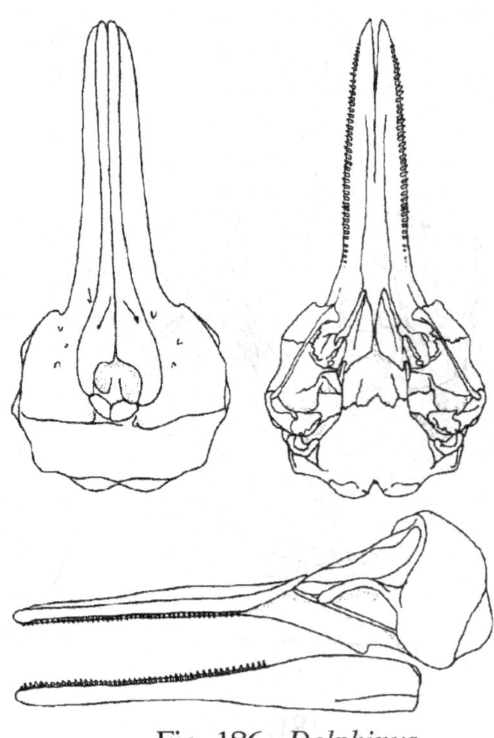

Fig. 186. *Delphinus*
Greatest length of skull 480mm

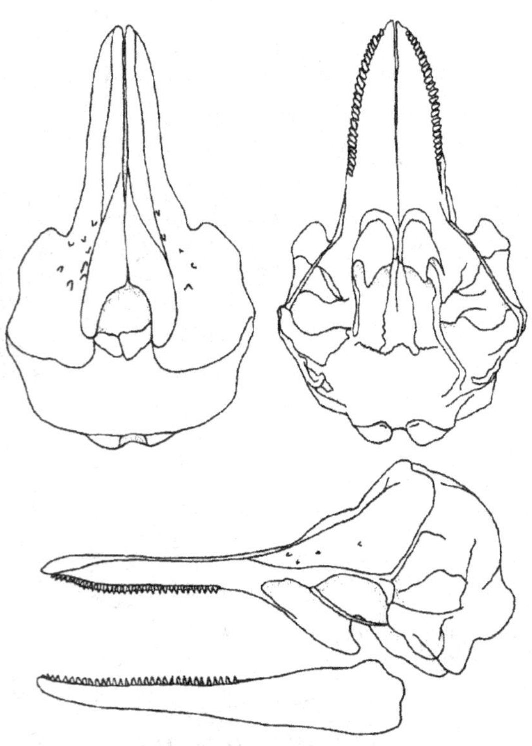

Fig. 187. *Peponocephala*
Greatest length of skull 415mm

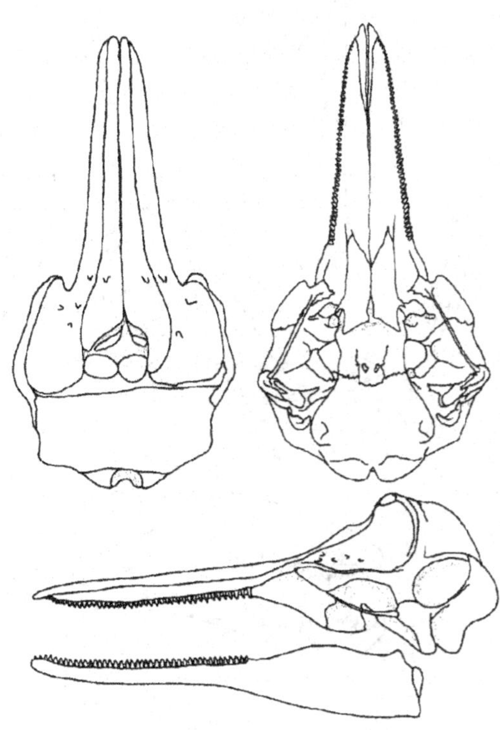

Fig. 188. *Stenella*
Greatest length of skull 435mm

-Rostral portion of premaxillae flat ................................................................ 26

26.    -Teeth small and acute, 43 or more in each row above
       and below ................................................................ *Lissodelphis* (Fig. 189)
       -Teeth larger, 26 or less in each row ................................ *Tursiops* (Fig. 190)

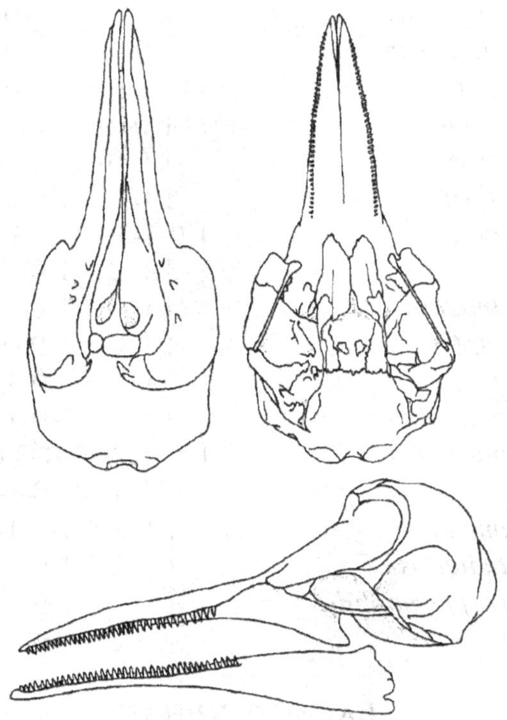

Fig. 189. *Lissodelphis*
Greatest length of skull 440mm

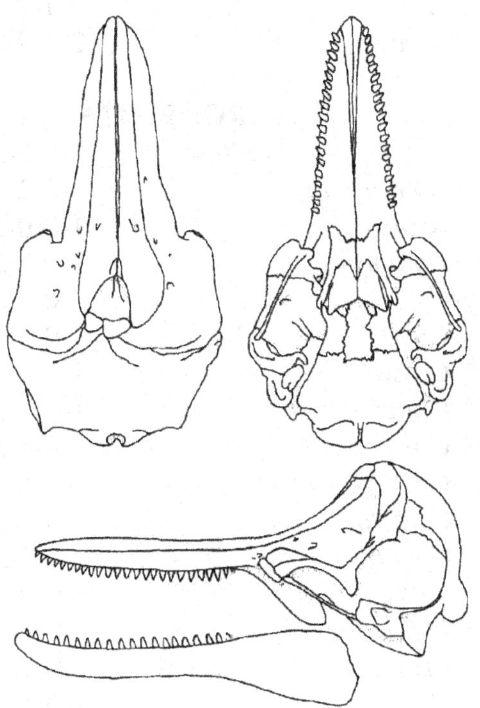

Fig. 190. *Tursiops*
Greatest length of skull 485mm

# DENTAL FORMULAS OF REPRESENTED GENERA

## DIDELPHIMORPHIA

| | |
|---|---|
| *Didelphis* | 5-1-3-4/4-1-3-4 |

## SIRENIA

| | |
|---|---|
| *Trichechus* | 0-0-6/0-0-6 |

## CINGULATA

| | |
|---|---|
| *Dasypus* | 0-0-7/0-0-7 or |
| | 0-0-8/0-0-8 |

## PRIMATES

| | |
|---|---|
| *Chlorocebus** | 2-1-2-3/2-1-2-3 |
| *Homo** | 2-1-2-3/2-1-2-3 |
| *Macaca** | 2-1-2-3/2-1-2-3 |

## RODENTIA

| | |
|---|---|
| *Ammospermophilus* | 1-0-2-3/1-0-1-3 |
| *Aplodontia* | 1-0-2-3/1-0-1-3 |
| *Arborimus* | 1-0-0-3/1-0-0-3 |
| *Baiomys* | 1-0-0-3/1-0-0-3 |
| *Callospermophilus* | 1-0-2-3/1-0-1-3 |
| *Castor* | 1-0-1-3/1-0-1-3 |
| *Cavia** | 1-0-1-3/1-0-1-3 |
| *Chaetodipus* | 1-0-1-3/1-0-1-3 |
| *Chinchilla** | 1-0-1-3/1-0-1-3 |
| *Cratogeomys* | 1-0-1-3/1-0-1-3 |
| *Cricetus** | 1-0-0-3/1-0-0-3 |
| *Cynomys* | 1-0-2-3/1-0-1-3 |
| *Dicrostonyx* | 1-0-0-3/1-0-0-3 |
| *Dipodomys* | 1-0-1-3/1-0-1-3 |
| *Erethizon* | 1-0-1-3/1-0-1-3 |
| *Geomys* | 1-0-1-3/1-0-1-3 |
| *Glaucomys* | 1-0-2-3/1-0-1-3 |
| *Ictodomys* | 1-0-2-3/1-0-1-3 |
| *Lemmiscus* | 1-0-0-3/1-0-0-3 |
| *Lemmus* | 1-0-0-3/1-0-0-3 |
| *Liomys* | 1-0-1-3/1-0-1-3 |
| *Marmota* | 1-0-2-3/1-0-1-3 |
| *Meriones** | 1-0-0-3/1-0-0-3 |
| *Mesocricetus** | 1-0-0-3/1-0-0-3 |
| *Microdipodops* | 1-0-1-3/1-0-1-3 |
| *Microtus* | 1-0-0-3/1-0-0-3 |
| *Mictomys* | 1-0-0-3/1-0-0-3 |
| *Mus** | 1-0-0-3/1-0-0-3 |
| *Myocastor** | 1-0-1-3/1-0-1-3 |
| *Myodes* | 1-0-0-3/1-0-0-3 |
| *Napaeozapus* | 1-0-0-3/1-0-0-3 |
| *Neofiber* | 1-0-0-3/1-0-0-3 |
| *Neotoma* | 1-0-0-3/1-0-0-3 |
| *Ochrotomys* | 1-0-0-3/1-0-0-3 |
| *Ondatra* | 1-0-0-3/1-0-0-3 |
| *Onychomys* | 1-0-0-3/1-0-0-3 |
| *Oryzomys* | 1-0-0-3/1-0-0-3 |
| *Otospermophilus* | 1-0-2-3/1-0-1-3 |
| *Perognathus* | 1-0-1-3/1-0-1-3 |
| *Peromyscus* | 1-0-0-3/1-0-0-3 |
| *Phenacomys* | 1-0-0-3/1-0-0-3 |
| *Podomys* | 1-0-0-3/1-0-0-3 |
| *Poliocitellus* | 1-0-2-3/1-0-1-3 |
| *Rattus** | 1-0-0-3/1-0-0-3 |
| *Reithrodontomys* | 1-0-0-3/1-0-0-3 |
| *Sciurus* | 1-0-1-3/1-0-1-3 or |
| | 1-0-2-3/1-0-1-3 |
| *Sigmodon* | 1-0-0-3/1-0-0-3 |
| *Synaptomys* | 1-0-0-3/1-0-0-3 |
| *Tamias* | 1-0-1-3/1-0-1-3 |
| *Tamias* (*Eutamias*) | 1-0-2-3/1-0-1-3 |
| *Tamiasciurus* | 1-0-2-3/1-0-1-3 or |
| | 1-0-1-3/1-0-1-3 |
| *Thomomys* | 1-0-1-3/1-0-1-3 |
| *Urocitellus* | 1-0-2-3/1-0-1-3 |
| *Xerospermophilus* | 1-0-2-3/1-0-1-3 |
| *Zapus* | 1-0-1-3/1-0-0-3 |

## LAGOMORPHA

| | |
|---|---|
| *Brachylagus* | 2-0-3-3/1-0-2-3 |
| *Lepus* | 2-0-3-3/1-0-2-3 |
| *Ochotona* | 2-0-3-2/1-0-2-3 |
| *Oryctolagus** | 2-0-3-3/1-0-2-3 |
| *Sylvilagus* | 2-0-3-3/1-0-2-3 |

## SORICOMORPHA

| | |
|---|---|
| *Blarina* | 3-1-3-3/1-1-1-3 |
| *Condylura* | 3-1-4-3/3-1-4-3 |
| *Cryptotis* | 3-1-2-3/1-1-1-3 |
| *Neurotrichus* | 3-1-2-3/3-1-2-3 |

| Notiosorex | 3-1-1-3/2-0-1-3 |
| Parascalops | 3-1-4-3/3-1-4-3 |
| Scalopus | 3-1-3-3/2-0-3-3 |
| Scapanus | 3-1-4-3/3-1-4-3 |
| Sorex | 3-1-3-3/1-1-1-3 |
| Sorex (Microsorex) | 3-1-1-3/1-1-1-3 |

## CHIROPTERA

| Antrozous | 1-1-1-3/2-1-2-3 |
| Artibeus | 2-1-2-2/2-1-2-3 |
| Choeronycteris | 2-1-2-3/0-1-3-3 |
| Corynorhynus | 2-1-2-3/3-1-3-3 |
| Diphylla | 2-1-1-2/2-1-2-2 |
| Eptesicus | 2-1-1-3/3-1-2-3 |
| Euderma | 2-1-2-3/3-1-2-3 |
| Eumops | 1-1-2-3/2-1-2-3 |
| Idionycteris | 2-1-2-3/3-1-3-3 |
| Lasionycteris | 2-1-2-3/3-1-3-3 |
| Lasiurus (Dasypterus) | 1-1-1-3/3-1-2-3 |
| Lasiurus (Lasiurus) | 1-1-2-3/3-1-2-3 |
| Leptonycteris | 2-1-2-2/2-1-3-2 |
| Macrotus | 2-1-2-3/2-1-3-3 |
| Molossus | 1-1-1-3/1-1-2-3 |
| Mormoops | 2-1-2-3/2-1-3-3 |
| Myotis | 2-1-3-3/3-1-3-3 |
| Nycticeius | 1-1-1-3/3-1-2-3 |
| Nyctinomops | 1-1-2-3/2-1-2-3 |
| Parastrellus | 2-1-2-3/3-1-2-3 |
| Perimyotis | 2-1-2-3/3-1-2-3 |
| Tadarida | 1-1-2-3/3-1-2-3 |

## CARNIVORA

| Alopex | 3-1-4-2/3-1-4-3 |
| Arctocephalus | 3-1-4-2/2-1-4-1 |
| Bassariscus | 3-1-4-2/3-1-4-2 |
| Callorhinus | 3-1-4-2/2-1-4-1 |
| Canis | 3-1-4-2/3-1-4-3 |
| Conepatus | 3-1-2-1/3-1-3-2 |
| Cystophora | 2-1-4-1/1-1-4-1 |
| Enhydra | 3-1-3-1/2-1-3-2 |
| Erignathus | 3-1-4-1/2-1-4-1 |
| Eumetopias | 3-1-4-1/2-1-4-1 |
| Felis | 3-1-3-1/3-1-2-1 |
| Gulo | 3-1-4-1/3-1-4-2 |
| Halichoerus | 3-1-4-1/2-1-4-1 |
| Leopardus | 3-1-3-1/3-1-2-1 |

| Lontra | 3-1-4-1/3-1-3-2 |
| Lynx | 3-1-2-1/3-1-2-1 |
| Martes | 3-1-4-1/3-1-4-2 |
| Mephitis | 3-1-3-1/3-1-3-2 |
| Mirounga | 2-1-4-1/1-1-4-1 |
| Monachus | 2-1-4-1/2-1-4-1 |
| Mustela | 3-1-3-1/3-1-3-2 |
| Nasua | 3-1-4-2/3-1-4-2 |
| Neovison | 3-1-3-1/3-1-3-2 |
| Odobenus | 1-1-3-0/0-1-3-0 |
| Panthera | 3-1-3-1/3-1-2-1 |
| Phoca | 3-1-4-1/2-1-4-1 |
| Procyon | 3-1-4-2/3-1-4-2 |
| Puma | 3-1-3-1/3-1-2-1 |
| Spilogale | 3-1-3-1/3-1-3-2 |
| Taxidea | 3-1-3-1/3-1-3-2 |
| Urocyon | 3-1-4-2/3-1-4-3 |
| Ursus | 3-1-4-2/3-1-4-3 |
| Vulpes | 3-1-4-2/3-1-4-3 |
| Zalophus | 3-1-4-1/2-1-4-1 or<br>3-1-4-2/2-1-4-1 |

## PERISSODACTYLA

| Equus* | 3-0-3-3/3-0-3-3 to<br>3-1-3-3/3-1-3-3 (canines variable) |

## ARTIODACTYLA

| Alces | 0-0-3-3/3-1-3-3 |
| Ammotragus* | 0-0-3-3/3-1-3-3 |
| Antilocapra | 0-0-3-3/3-1-3-3 |
| Antilope* | 0-0-3-3/3-1-2-3 |
| Axis* | 0-0-3-3/3-1-3-3 |
| Bison | 0-0-3-3/3-1-3-3 |
| Bos | 0-0-3-3/3-1-3-3 |
| Boselaphus* | 0-0-3-3/3-1-3-3 |
| Camelus* | 1-1-3-3/3-1-2-3 |
| Capra* | 0-0-3-3/3-1-3-3 |
| Cervus | 0-1-3-3/3-1-3-3 |
| Dama* | 0-0-3-3/3-1-3-3 |
| Lama* | 1-1-2-3/3-1-1-3 |
| Odocoileus | 0-0-3-3/3-1-3-3 |
| Oreamnos | 0-0-3-3/3-1-3-3 |
| Oryx* | 0-0-3-3/3-1-3-3 |
| Ovibos | 0-0-3-3/3-1-3-3 |
| Ovis | 0-0-3-3/3-1-3-3 |
| Pecari | 2-1-3-3/3-1-3-3 |

| *Rangifer* | 0-0-3-3/3-1-3-3 or |
| | 0-1-3-3/3-1-3-3 |
| *Sus** | 3-1-4-3/3-1-4-3 |
| *Taurotragus** | 0-0-3-3/3-1-3-3 |

## CETACEA

Genera in Suborder Mysticeti with baleen, lack teeth after birth.

Genera in Odontoceti with one or more pairs of teeth, total number of teeth variable.

# SELECTED REFERENCES

Abramov, A.V. 2000. A taxonomic review of the genus *Mustela* (Mammalia, Carnivora). Zoosystematica Rossica, 8:357-364.

Allen, G. M. 1942. Extinct and vanishing mammals of the western hemisphere with the marine species of all the oceans. Special Pub. No. 11, Amer. Commit. Int. Wild Life Protection, Washington D.C., 620pp. (1972 reprint by Cooper Square, New York.)

Allen, J. A. 1880. History of North American pinnipeds: A monograph of the walruses, sea-lions, sea-bears and seals of North America. U. S. Geol. Geog. Surv. Terr. (Hayden Survey) Misc. Publ. 12, 785pp. (1974 reprint by Arno Press, New York.)

Ammerman, L. K., C. L. Hice, and D. J. Schmidly. 2012. Bats of Texas. Texas A&M Press, College Station, 328pp.

Anderson, R. M. 1947. Catalogue of Canadian Recent mammals. Nat. Mus. Canada Bull., Bio. Series, 102:1-238.

Anderson, S., and J. K. Jones, Jr. (eds.). 1984. Orders and families of recent mammals of the world. John Wiley & Sons, New York, 686pp.

Armstrong, D. J., Fitzgerald, J. P., and C. A. Meaney. 2011. Mammals of Colorado. Univ. Press of Colorado, Boulder, 620pp.

Bailey, V. 1936. The mammals and life zones of Oregon. North Amer. Fauna, 55:1-416.

Burt, W. H. 1946. The mammals of Michigan. Univ. Mich. Press, Ann Arbor, 288pp.

Baker, A. J., J. L. Eger, R. L. Peterson, and T. H. Manning. 1983. Geographic variation and taxonomy of arctic hares. Acta Zool. Fennica, 174:45-48.

Baker, R. J., J. K. Jones, Jr., and D. C. Carter (eds.). 1976. Biology of bats of the New World family Phyllostomatidae. Special Publ., The Museum, Texas Tech Univ. Press, 10:1-218.

Banfield, A. W. F. 1974. The mammals of Canada. Univ. Toronto Press, Toronto, 438pp.

Barbour, R. W., and W. H. Davis. 1969. Bats of America. Univ. Kentucky Press, Lexington, 286pp.

Barnes, L. G. 1985. Evolution, taxonomy and antitropical distribution of the porpoises (Phocoenidae, Mammalia). Marine Mammal Sci., 1(2):149-163.

Bee, J. W., and E. R. Hall. 1956. Mammals of northern Alaska on the Arctic Slope. Misc. Pub., Mus. Nat. Hist., Univ. Kansas, 8:1-309.

Bellinger, M. R., S. M. Haig, E. D. Forsman, and T. D. Mullins. 2005. Taxonomic relationships among *Phenacomys* voles as inferred by cytochrome *b*. J. Mammal., 86:201-210.

Bonner, W. N. 1990. The natural history of seals. Facts on File, New York, 196pp.

Boorer, M. 1971. Mammals of the world. Grosset and Dunlap, New York, 156pp.

Broadbooks, H. E. 1965. Ecology and distribution of the pikas of Washington and Alaska. Amer. Midl. Nat., 73:299-335.

Brown, L. N. 1997. A guide to the mammals of the Southeastern United States. The Univ. Tennessee Press, Knoxville, 236pp.

Bryant, M. D. 1945. Phylogeny of Nearctic Sciuridae. Amer. Midl. Nat., 33:257-390.

Bubenik, G. A., and A. B. Bubenik (eds.). 1990. Horns, pronghorns, and antlers. Springer-Verlag, New York, 562pp.

Bueler, L. E. 1973. Wild dogs of the world. Stein & Day, New York, 274 pp.

Burns, J. J., and F. H. Fay. 1970. Comparative morphology of the skull of the ribbon seal, *Histriophoca fasciata*, with remarks on systematics of Phocidae. J. Zool. (London), 161:363-394.

Burt, W. H., and R. P. Grossenheider. 1976. A field guide to the mammals of America north of Mexico. 3rd ed. Peterson Field Guide Series No. 5, Houghton Mifflin Co., Boston, 289pp.

Caire, W., J. D. Tyler, B. P. Glass, and M. A. Mares. 1989. Mammals of Oklahoma. Univ. Okla. Press, Norman, 567pp.

Carleton, M. D. 1980. Phylogenetic relationships in neotomine-peromyscine rodents (Muroidea) and a reappraisal of the dichotomy within New World Cricetinae. Misc. Publ., Mus. Zool., Univ. Mich., 157:1-146.

Carwardine, M. 1995. Whales, dolphins, and porpoises. Eyewitness Handbooks, Dorling Kindersley, New York, 256pp.

Chapman, J. A., and G. A. Feldhamer (eds.). 1982. Wild mammals of North America: Biology, management, and economics. Johns Hopkins Univ. Press, Baltimore, 1147pp.

Chapman, J. A., and J. E. C. Flux (eds.). 1990. Rabbits, hares and pikas. I. U. C. N., Gland, Switzerland, 168 pp.

Chapman, J. A., K. R. Dixon, W. Lopez-Forment, and D. E. Wilson. 1983. The New World jackrabbits and hares (genus *Lepus*) 1. Taxonomic history and population status. Acta

Zool. Fennica, 174:49-51.

Cockrum, E. L., and Y. Petryszyn. 1992. Mammals of the southwestern United States and northwestern Mexico. Treasure Chest Publ., Tucson, 192pp.

Corbet, G. B. 1988. The family Erinaceidae: A synthesis of its taxonomy, phylogeny, ecology and zoogeography. Mammal Review, 18:117-172.

Coues, E. 1877. Fur-bearing animals: A monograph of North American Mustelidae etc. Misc. Publ. U. S. Geol. Surv. Terr., 348pp.

Coues, E., and J. A. Allen. 1877. Monographs of North American Rodentia. Misc. Publ. U. S. Geol. Surv. vol. 11, 1091pp.

Cowan, I. McT. 1940. Distribution and variation in the native sheep of North America. Amer. Midl. Nat., 24:505-580.

Cowan, I. McT., and C. J. Guiguet. 1965. The mammals of British Columbia. 3rd ed. B. C. Prov. Mus., Handbook 11:1-414.

Cox, P. G., and L. Hautier (eds.) 2015. Evolution of the Rodents: Volume 5: Advances in Phylogeny, Functional Morphology and Development. Cambridge Studies in Morphology and Molecules: New Paradigms in Evolutionary Biology, Cambridge University Press, 624pp.

Decker, D. M., and W. C. Wozencraft. 1991. Phylogenetic analysis of Recent procyonid genera. J. Mammal., 72:42-55.

Diersing, V. A. 1980. Systematics and evolution of the pygmy shrews (subgenus *Microsorex*) of North America. J. Mammal., 61:76-101.

Dixon, K. R., J. A. Chapman, G. R. Willner, D. E. Wilson, and W. Lopez-Forment. 1983. The New World jackrabbits and hares (genus *Lepus*) 2. Numerical taxonomic analysis. Acta Zool. Fennica, 174:53-56.

Dobson, G. E. 1882. A monograph of the Insectivora, systematic and anatomical, Part 1: Including the families Erinaceidae, Centetidae, and Solenodontidae. John Van Voorst, London, 96pp.

Dolan, J. M. 1988. A deer of many lands - a guide to the subspecies of the red deer *Cervus elaphus*. L. Zoonooz, 62(10):4-34.

Domning, D. P. 1978. Sirenian evolution in the North Pacific Ocean. Univ. Calif. Publ. Geol. Sci., 118:1-176.

Dragoo, J. W., J. R. Choate, T. L. Yates, and T. P. O'Farrell. 1990. Evolutionary and taxonomic relationships among North American arid-land foxes. J. Mammal., 71:318-332.

East, R. (ed.). 1988. Antelopes. Global survey and regional action plans. Part 1. East and northeast Africa. I. U. C. N., Gland, Switzerland, 96pp.

______. 1989. Antelopes. Global survey and regional action plans. Part 2. Southern and south-central Africa. I. U. C. N., Gland, Switzerland, 96pp.

______. 1990. Antelopes. Global survey and regional action plans. Part 3. West and central Africa. I. U. C. N., Gland, Switzerland, 171pp.

Elbroch, M. 2006. Animal skulls: a guide to North American species. Stackpole Books, Mechanicsburg, Pennsylvania, 727pp.

Ellerman, J. R. 1940. The families and genera of living rodents. Vol. I. Rodents other than Muridae. British Mus. (Nat. Hist.), London, 689pp.

______. 1941. The families and genera of living rodents. Vol. II. Family Muridae. British Mus. (Nat. Hist.), London, 690pp.

Elliot, D. G. 1901. A synopsis of the mammals of North America and the adjacent seas. Field Columb. Mus. Zool. Ser. vol. 2, 471pp.

______. 1904. The land and sea mammals of Middle America and the West Indies. Field Mus. Publ. 95, 2 parts, 850pp.

Ellis, L. S., and L. R. Maxson. 1979. Evolution of the chipmunk genera *Eutamias* and *Tamias*. J. Mammal., 60:331-334.

Ellis, R. 1980. The book of whales. Alfred A. Knopf, New York, 202pp.

Ewer, R. F. 1973. The carnivores. Cornell Univ. Press, Ithaca, New York, 494pp.

Feldhamer, G. A., L. C. Drickamer, J. F. Merritt, S. H. Vessey, and C. Krajewski. 2008. Mammalogy: adaptations, diversity, ecology (3rd ed.). Johns Hopkins University Press, Baltimore, Maryland, 672pp.

Feldhammer, G. A., B. C. Thompson, and J. A. Chapman (eds.). 2003. Wild Mammals of North America: Biology, Management, and Conservation (2nd ed.). Johns Hopkins University Press, Baltimore, Maryland, 1232pp.

Findley, J. S., A. H. Harris, D. E. Wilson, and C. Jones. 1975. Mammals of New Mexico. Univ. New Mexico Press, Albuquerque, 360pp.

Finley, R. B. 1958. The wood rats of Colorado: Distribution and ecology. Univ. Kansas Publ. Mus. Nat. Hist., 10:213-552.

Fisler, G. F. 1970. Keys to identification of the orders and families of living mammals of the world. Sci. Series 25, Zool. No. 12, 29pp.

Flerov, K. K. 1960. Fauna of the U.S.S.R. and adjacent countries vol. I(2): Musk deer and deer. Israel Prog. Sci. Trans., Jerusalem, 257pp.

Foresman, K. R. 2001. Key to the mammals of Montana. The Bookstore, University of Montana, Missoula, 92pp.

Forsythe, A. 1985. Mammals of the American north. Camden House, Ontario, 351pp.

Foster-Turley, P., S. Macdonald, and C. Mason (eds.). 1990. Otters: An action plan for their conservation. I. U. C. N., Gland, Switzerland, 126pp.

Freeman, P. W. 1981. A multivariate study of the family Molossidae (Mammalia: Chiroptera): morphology, ecology, evolution. Fieldiana, Zool. New Series, 7:1-173.

Gardner, A. L. 1973. The systematics of the genus *Didelphis* (Marsupialia: Didelphidae) in North and Middle America. Special Publ., The Museum, Texas Tech Univ., 4:1-81.

Gauthier-Pilters, H., and A. Innis Dagg. 1981. The camel: Its evolution, ecology, behavior and relationships to man. Univ. Chicago Press, Chicago, 208pp.

Genoways, H. H. 1973. Systematics and evolutionary relationships of spiny pocket mice, genus *Liomys*. Special Publ., The Museum, Texas Tech Univ., 5:1-368.

George, S. B. 1986. Evolution and historical biogeography of soricine shrews. Systematic Zoology, 35:153-162.

Ginsberg, J. R., and D. W. Macdonald. 1990. Foxes, wolves, jackals, and dogs: An action plan for the conservation of canids. I. U. C. N., Gland, Switzerland, 116pp.

Gittleman, J. L. (ed.). 1989. Carnivore behavior, ecology and evolution. Cornell Univ. Press, Ithaca, 620pp.

Gorman, M. L., and R. D. Stone. 1990. The natural history of moles. Comstock, Ithaca, 160pp.

Groves, C. P. 1974. Horses, asses and zebras in the wild. David and Charles, Newton Abbot and London, 192pp.

______. 1982. Cranial and dental characteristics in the systematics of Old World Felidae. Carnivore, 5(2):28-39.

Groves, C. P., and D. P. Willoughby. 1981. Studies on the taxonomy and phylogeny of the genus *Equus*. 1. Subgeneric classification of the recent species. Mammalia, 45:321-354.

Hafner, J. C., and M. S. Hafner. 1983. Evolutionary relationships of heteromyid rodents. Great Basin Nat. Mem., 7:3-29.

Haley, D. (ed.). 1986. Marine mammals of eastern North Pacific and Arctic waters. 2nd ed. Pacific Search Press, Seattle, 256pp.

Hall, E. R. 1946. Mammals of Nevada. Univ. Calif. Press, Berkeley, 710pp.

______. 1951. American weasels. Univ. Kansas Publ. Mus. Nat. Hist., 4:1-466.

______. 1955. Handbook of mammals of Kansas. Misc. Publ. Mus. Nat. Hist., Univ. Kansas, 7:1-303.

______. 1981. Mammals of North America. 2nd ed., 2 vols., John Wiley & Sons, New York, 1181pp.

Hall, E. R., and E. L. Cockrum. 1953. A synopsis of the North American Microtine rodents. Univ. Kansas Publ., Mus. Nat. Hist., 5:373-498.

Hall, E. R., and J. K. Jones, Jr. 1961. North American yellow bats, "*Dasypterus*", and a list of the named kinds of the genus *Lasiurus* Gray. Univ. Kansas Publ., Mus. Nat. Hist., 14:73-98.

Hamilton, W. J., and J. O. Whitaker. 1979. Mammals of the eastern United States. Cornell Univ. Press, Ithaca, 346pp.

Handley, C. O., Jr. 1959. A revision of American bats of the genera *Euderma* and *Plecotus*. Proc. U. S. Nat. Mus., 110:95-246.

Harris, C. J. 1968. Otters: A study of the Recent Lutrinae. Weidenfeld and Nicolson, London, 397pp.

Harrison, R. J., and J. E. King. 1980. Marine mammals. 2nd ed. Hutchinson & Co., London, 192pp.

Helgen, K. M., F. R. Cole, L. E. Helgen, and D. E. Wilson. 2009. Generic revision in the Holarctic ground squirrel genus *Spermophilus*. J. Mammal., 90:270-305.

Hershkovitz, P. 1966. Catalog of living whales. Bull. U. S. Nat. Mus., 246:1-259.

Heyning, J. E. 1989. Comparative facial anatomy of beaked whales (Ziphiidae) and a systematic revision among the families of extant Odontoceti. Contrib. Sci., Nat. Hist. Mus., Los Angeles Co., 405:1-64.

Hillson, S. 2005. Teeth. Cambridge Univ. Press, Cambridge, United Kingdom, 373pp.

Hoffmeister, D. F. 1986. Mammals of Arizona. Univ. Arizona Press, Tucson, 602pp.

Hollister, N. 1911. A systematic synopsis of the muskrats. North Amer. Fauna, 32:1-47.

Honacki, J. H., K. E. Kinman, and J. W. Koeppl (eds.). 1982. Mammal species of the world: A taxonomic and geographic reference. Allen Press, Inc., and Assoc. Syst. Coll.,

Lawrence, Kansas, 694pp.

Hoofer, S. R., R. A. Van Den Bussche, and I. HoráČek. 2006. Generic status of the American Pipistrelles (Vespertilionidae) with description of a new genus. J. Mammal., 87:981-992.

Hooper, E. T. 1957. Dental patterns in mice of the genus *Peromyscus*. Misc. Publ., Mus. Zool., Univ. Mich., 99:1-59.

Hooper, E. T., and B. S. Hart. 1962. A synopsis of Recent North American microtine rodents. Misc. Publ., Mus. Zool., Univ. Mich., 120:1-68.

Howell, A. H. 1938. Revision of the North American ground squirrels, with a classification of the North American Sciuridae. North Amer. Fauna, 56:1-256.

Hunt, R. M., Jr. 1974. The auditory bulla in Carnivora: An anatomical basis for reappraisal of carnivore evolution. J. Morph., 143:21-76.

Hutterer, R. 1985. Anatomical adaptations of shrews. Mammal Review, 15:43-55.

Ingles, L. G. 1965. Mammals of the Pacific states: California, Oregon, and Washington. Stanford Univ. Press, Stanford, 506pp.

Jones, J. K., Jr., and R. W. Manning. 1992. Illustrated key to skulls of genera of North American land mammals. Texas Tech Univ. Press, Lubbock, 75pp.

Jones, J. K., Jr., R. S. Hoffmann, D. W. Rice, C. Jones, R. J. Baker, and M. D. Engstrom. 1992. Revised checklist of North American mammals north of Mexico, 1991. Occ. Papers, The Museum, Texas Tech Univ., 146:1-23.

Joysey, K. A., and T. S. Kemp (eds.). 1972. Studies in vertebrate evolution. Oliver and Boyd, Edinburgh, 284pp.

Junge, J. A., and R. S. Hoffmann. 1981. An annotated key to the long-tailed shrews (genus Sorex) of the United States and Canada, with notes on Middle American Sorex. Occ. Papers Mus. Nat. Hist., Univ. Kansas, 94:1-48.

Kasuya, T. 1973. Systematic consideration of recent toothed whales based on the morphology of the tympano-periotic bone. Sci. Reports, Whales Res. Inst., Tokyo, 25:1-103.

Katona, S. K., V. Rough, and D. T. Richardson. 1993. A field guide to whales, porpoises and seals from Cape Cod to Newfoundland. Smithsonian Inst. Press, Washington D.C., 316pp.

Kenyon, K. W. 1969. The sea otter in the eastern Pacific Ocean. North Amer. Fauna, 68:1-352.

Keogh, H. J. 1985. A photographic reference system based on the cuticular scale patterns and grooves of the hair of 44 species of southern African Cricetidae and Muridae. South African J. Wildl. Res., 15:109-159.

Kesner, M. H. 1980. Functional morphology of the masticatory musculature of the rodent subfamily Microtinae. J. Morph., 165:205-222.

King, J. A. (ed.). 1968. Biology of *Peromyscus* (Rodentia). Spec. Publ., Amer. Soc. Mammal., 2:1-593.

King, J. E. 1954. The otariid seals of the Pacific coast of America. Bull. British Mus. (Nat. Hist.), Zool. Series, 2(10):311-337.

______. 1966. Relationships of the hooded and elephant seals (genera *Cystophora* and *Mirounga*). J. Zool. (London), 148:385-398.

______. 1983. Seals of the world. 2nd ed. British Mus. and Cornell Univ. Press, Ithaca, 240pp.

Kirkland, G. L., Jr., and J. N. Layne (eds.). 1989. Advances in the study of *Peromyscus* (Rodentia). Texas Tech Univ. Press, Lubbock, 367pp.

Kurtu, A. 1995. Mammals of the Great Lakes Region. Univ. Mich. Press, Ann Arbor, 376pp.

Lawler, T. E. 1976. Handbook to the orders and families of living mammals. Mad River Press, Eureka, 244pp.

Leatherwood, S., and R. R. Reeves (eds.). 1990. The bottlenose dolphin. Academic Press, New York, 653pp.

Lechleitner, R. R. 1969. Wild mammals of Colorado: Their appearance, habits, distribution and abundance. Pruett Publ. Co., Colo., 254pp.

Levenson, H., R. S. Hoffmann, C. F. Nadler, L. Deutsch, and S. D. Freeman. 1985. Systematics of the Holarctic chipmunks. J. Mammal., 66:219-242.

Lever, C. 1985. Naturalized mammals of the World. Longman, London, 487pp.

Linzey, D. W. 1998. The mammals of Virginia. McDonald & Woodward Publ. Co., Blacksburg, Virginia, 459pp.

Long, C. A., and C. A. Killingley. 1983. The badgers of the world. Charles C. Thomas, Springfield, 404pp.

Lowery, G. H., Jr. 1974. The mammals of Louisiana and its adjacent waters. Louisiana State Univ. Press, Baton Rouge, 565pp.

Luckett, W. P., and J. L. Hartenberger (eds.). 1985. Evolutionary relationships among rodents, a multidisciplinary analysis. Plenum Press, New York, 721pp.

Macdonald, D. (ed.). 1984. The encyclopedia of mammals. Facts on File Publ., New York, 895pp.

Mansfield, A. W. 1967. Seals of Arctic and Eastern Canada. 2nd ed. Bull. 137, Fisheries Research Board of Canada, Queen's Printer, Ottawa, 35pp.

Martin, L. D. 1980. The early evolution of the Cricetidae in North America. Univ. Kansas Paleo. Contrib., 102:1-42.

Martin, R. E., R. H. Pine, and A. F. DeBlase. 2011. A manual of mammalogy: with keys to families of the world (3$^{rd}$ spi. ed.). Waveland Press, Long Grove, Illinois, 333pp.

McDonald, J. N. 1981. North American bison, their classification and evolution. Univ. Calif. Press, Berkeley, 316pp.

Mead, J. G. 1975. Anatomy of the external nasal passages and facial complex in the Delphinidae (Mammalia: Cetacea). Smithsonian Contrib. Zool., 207:1-72.

Mearns, E. A. 1907. Mammals of the Mexican Boundary: Part 1. U. S. Nat. Mus. Bull. 56, 530pp.

Meester, J., and H. W. Setzer (eds.). 1977. The mammals of Africa: An identification manual. Smithsonian Inst. Press, Washington D.C., not continuously paginated.

Miller, G. S., Jr. 1925. The telescoping of the Cetacean skull. Smithsonian Misc. Coll., 76(5):1-70.

Miller, G. S., Jr., and R. Kellogg. 1955. List of North American Recent mammals. Bull. U. S. Nat. Mus., 205:1-954.

Miller, G. S., Jr., and J. A. G. Rehn. 1901. Systematic results of the study of North American land mammals to the close of the year 1900. Proc. Boston Soc. Nat. Hist., 30:1-352.

Montgomery, G. G. (ed.). 1985. The evolution and ecology of armadillos, sloths, and vermilinguas. Smithsonian Inst. Press, Washington D.C., 451pp.

Moore, J. C. 1959. Relationships among living squirrels of the Sciurinae. Bull. Amer. Mus. Nat. Hist., 118:157-206.

______. 1968. Relationships among the living genera of beaked whales with classifications, diagnoses and keys. Fieldiana, Zool., 53(4):209-298.

Murie, J. O., and G. R. Michener (eds.). 1984. The biology of ground-dwelling squirrels. Univ. Nebr. Press, Lincoln, 459pp.

Murie, O. J. 1935. Alaska-Yukon caribou. North Amer. Fauna, 54:1-93.

Nelson, E. W. 1909. The rabbits of North America. North Amer. Fauna, 29:1-314.

Nelson, E. W., and E. A. Goldman. 1933. Revision of the jaguars. J. Mammal., 14(3):221-240.

Norman, J. N., and F. C. Fraser. 1938. Giant fishes, whales, and dolphins. Putnam, London, 361pp.

Norris, K. S. (ed.). 1966. Whales, dolphins and porpoises. Univ. Calif. Press, Berkeley, 789pp.

Nowak, R. M. 1991. Walker's mammals of the world. 5th ed. 2 vols. Johns Hopkins Univ. Press, Baltimore, 1629pp.

______. 1994. Walker's bats of the world. Johns Hopkins Univ. Press, Baltimore, 287pp.

Ognev, S. I. 1962. Mammals of Eastern Europe and northern Asia vol. 1: Insectivora and Chiroptera. Israel Prog. Sci. Transl., Jerusalem, 487pp.

______. 1962. Mammals of Eastern Europe and northern Asia vol. 2: Carnivora (Fissipedia). Israel Prog. Sci. Transl., Jerusalem, 590pp.

______. 1962. Mammals of the U.S.S.R. and adjacent countries vol. 3: Fissipedia and Pinnipedia. Israel Prog. Sci. Transl., Jerusalem, 641pp.

______. 1963. Mammals of the U.S.S.R. and adjacent countries vol. 5: Rodents (cont.). Israel Prog. Sci. Transl., Jerusalem, 662pp.

______. 1963. Mammals of the U.S.S.R. and adjacent countries vol. 6: Rodents (cont.). Israel Prog. Sci. Transl., Jerusalem, 508pp.

______. 1964. Mammals of the U.S.S.R. and adjacent countries vol. 7: Rodents. Israel Prog. Sci. Transl., Jerusalem, 626pp.

______. 1966. Mammals of the U.S.S.R. and adjacent countries vol. 4: Rodents (Mammals of eastern Europe and northern Asia). Israel Prog. Sci. Transl., Jerusalem, 429pp.

Parker, S. B. (ed.). 1990. Grzimek's encyclopedia of mammals. 5 vols. McGraw Hill, New York, not continuously paginated.

Pelton, M. R., J. W. Lentfer, and G. E. Folk (eds.). 1976. Bears: Their biology and management. Int. Union Cons. Nat., New Series, 40:1-467.

Peterson, R.L. 1966. The mammals of eastern Canada. Oxford Univ. Press, Toronto, 465pp.

Pike, G. C., and I. B. MacAskie. 1969. Marine mammals of British Columbia. Fish. Res. Board of Canada, Bull., 171:1-54.

Quay, W. B. 1954. The anatomy of the diastemal palate in microtine rodents. Misc. Publ., Mus. Zool., Univ. Mich., 86:1-41.

Reeves, R.R., B.S. Stewart, and S. Leatherwood. 1992. The Sierra Club handbook of seals and sirenians. Sierra Club Books, San Francisco, 359pp.

Reid, F. 2006. Peterson Field Guide to Mammals of North America (4th ed.). Houghton Mifflin Harcourt, Boston, Massachusetts, 608pp.

Ridgway, S. H., and R. Harrison (eds.). 1981. Handbook of marine mammals vol. 1: The walrus, sea lions, fur seals and sea otter. Academic Press, New York, 235pp.

______. 1981. Handbook of marine mammals vol. 2: The seals. Academic Press, New York, 359pp.

______. 1985. Handbook of marine mammals vol. 3: The sirenians and baleen whales. Academic Press, New York, 330pp.

______. 1989. Handbook of marine mammals vol. 4: River dolphins and the larger toothed whales. Academic Press, New York, 512pp.

______. 1991. Handbook of marine mammals vol. 5: Dolphins. Academic Press, New York, 416pp.

Riedman, M. 1990. The Pinnipeds: Seals, sea lions, and walruses. Univ. Calif. Press, Berkeley, 439pp.

Robovský, J., V. Říčánková, and J. Zrzavý. 2008. Phylogeny of Arvicolinae (Mammalia, Cricetidae): utility of morphological and molecular data sets in a recently radiating clade. Zoologica Scripta, 37:571-590.

Roest, A. I. 1991. A key-guide to mammal skulls and lower jaws. Mad River Press, Eureka, 39pp.

Scheffer, V. B. 1958. Seals, sea lions, and walruses: A review of the Pinnipedia. Stanford Univ. Press, Stanford, 179pp.

Schmidly, D. J. 1994. The mammals of Texas. The University of Texas Press, Austin, 501pp.

Sealander, J. A., and G. A. Heidt. 1990. Arkansas mammals: their natural history, classification, and distribution. Univ. Arkansas Press, Fayetteville. 308pp.

Shaughnessy, P. D., and F. H. Fay. 1977. A review of the taxonomy and nomenclature of north Pacific harbour seals. J. Zool. (London), 182:385-419.

Sisson, S., and J. D. Grossman. 1928. The anatomy of the domestic animals. 3rd ed. W. B. Saunders Co., Philadelphia, 972pp.

Skinner, J. D., and C. T. Chimimba. 2005. The mammals of the southern African subregion. 3$^{rd}$ ed. Cambridge University Press, Cape Town, South Africa, 814pp.

Slaughter, B. H., and D. W. Walton (eds.). 1970. About bats. Southern Methodist Univ. Press, Dallas, 339pp.

Teaford, M. F., M. M. Smith, and M. W. J. Ferguson (eds.). 2006. Development, function, and evolution of teeth. Cambridge Univ. Press, New York, 314pp.

Tedford, R. H. 1976. Relationship of pinnipeds to other carnivores (Mammalia). Syst. Zool., 25:363-374.

Trani, M. K., W. M. Ford, and B. R. Chapman (eds.) 2007. The land manager's guide to mammals of the South. The Nature Conservancy, Durham, North Carolina and U.S. Forest Service, Atlanta, Georgia, 546 p.

Tumlison, R., and M. E. Douglas. 1992. Parsimony analysis and the phylogeny of the Plecotine bats (Chiroptera: Vespertilionidae). J. Mammal., 73:276-285.

Unger, P. S. 2010. Mammal teeth. The Johns Hopkins Univ. Press, Baltimore, Maryland, 304pp.

van Zyll de Jong, C. G. 1972. A systematic review of the Nearctic and Neotropical river otters (genus *Lutra*, Mustelidae, Carnivora). Royal Ontario Mus., Life Sci., Contrib., 80:1-104.

______. 1983. Handbook of Canadian mammals. Part I. Marsupials and insectivores. Nat. Mus. Nat. Sci., Ottawa, 210pp.

______. 1986. A systematic study of Recent bison, with particular consideration of the wood bison (*Bison bison athabascae* Rhoads 1898). Publ. Nat. Sci., Nat. Mus. Nat. Sci., Canada, 6:1-69.

Vaughan, T. A., J. M. Ryan, and N. J. Czaplewski. 2010. Mammalogy (5th ed.). Jones & Bartlett Learning, Burlington, Maryland, 750pp.

Vinogradov, B. S., and A. I. Argiropulo. 1968. Fauna of the U.S.S.R.: Mammals: Key to rodents. Israel Prog. Sci. Transl., Jerusalem, 230pp.

Wahlert, J. H. 1985. Skull morphology and relationships of geomyoid rodents. Amer. Mus. Novitates, 2812:1-20.

Wemmer, C. M. (ed.). 1987. Biology and management of the Cervidae. Smithsonian Inst. Press, Washington D.C., 577pp.

Whitaker Jr., J. O., and W. J. Hamilton Jr. 1998. Mammals of the Eastern United States (3$^{rd}$ ed.). Comstock Publishing Associates, Cornell University Press, Ithaca, New York, 608pp.

Whitehead, G. K.  1972.  Deer of the world. Constable, London, 194pp.

Wilson, D. E., and D.  M. Reeder (eds.).  2005.  Mammal species of the world: A taxonomic and geographic reference. 3$^{rd}$ ed. Johns Hopkins University Press, 2,142 pp.

Youngman, P. M.  1975.  Mammals of the Yukon Territory. Nat. Mus. Nat. Sci., Ottawa, Publ. Zool., 10:1-192.

Zangerl, R. (ed.).  1968.  Comparative odontology. U. Chicago Press, Chicago, Illinois, 347pp. + 88 b&w and 8 color plates.

Zeveloff, S. I.  1988.  Mammals of the intermountain west. Univ. Utah Press, Salt Lake City, 365pp.